Scientific plant names derived from the surname of Pehr Forsskål

– their history, orthography, legitimacy, and current use

Abstract

Pehr Forsskål's posthumously published *Flora Aegyptiaco-Arabica* (1775) contained information about ca. 660 new species of plants, but was defectively edited. Later authors often misunderstood the text and provided more than 10% of Forsskål's new, but already named species with new scientific plant names derived from Forsskål's surname (eponyms). Seventy-three Forsskål eponyms have been proposed for plants with spelling often causing difficulties for non-Scandinavian scholars; the Forsskål eponyms have been called "an amazing series of orthographic variations", with 34 variants in original botanical publications and many more in secondary publications. Over one hundred Forsskål eponyms have been published in zoological literature, but are not further dealt with here. The current *International Code of Nomenclature for Algae, Fungi and Plants* (ICN, with amendments made by the XX Botanical Congress in Madrid, 2024), requests that the spelling of Forsskål's eponyms in the original publications must be retained, apart from corrections proscribed in ICN's Art. 60–61, thereby preventing standardisation based directly on the surname as Forsskål himself wrote it. Having discussed what to correct under the ICN and what not, we review the original eponyms and their correct orthography, after which the number of variants is reduced from 34 to 16. We also discuss if the eponyms are valid and legitimate under ICN. Of the eponyms based on Forsskål's surname, seven are not validly published, 30 are illegitimate, and 36 are legitimate (of which one was rejected by the XVIII International Botanical Congress and one currently proposed for rejection). Of the legitimate eponyms, 19 provide epithets for currently accepted names of taxa.

Keywords: Egypt; eponyms; *Flora Aegyptiaco-Arabica*; homonyms; misidentifications; orthographic variants; superfluous names; Yemen.

Resumé

Pehr Forsskåls posthumt udgivne *Flora Aegyptiaco-Arabica* (1775) indeholdt oplysninger om ca. 660 nye plantearter, men var mangelfuldt redigeret. Senere forfattere misforstod ofte teksten og forsynede mere end 10 % af Forsskåls nye, men allerede navngivne arter med nye videnskabelige navne, der var afledt af Forsskåls efternavn (eponymer). Der er foreslået 73 Forsskål-eponymer for planter, hvor stavemåden ofte volder vanskeligheder for ikke-skandinaviske forskere. Forsskål-eponymerne er blevet kaldt "en fantastisk serie af ortografiske variationer", med 34 varianter i originale botaniske publikationer og mange flere i sekundære publikationer. Over hundrede Forsskål-eponymer er blevet udgivet i zoologisk litteratur, men de behandles ikke yderligere her. Den nugældende *Internationale Nomenklaturkode for Alger, Svampe og Planter* (ICN, med ændringer foretaget af den XX. botaniske kongres i Madrid, 2024) kræver, at stavemåden af Forsskåls eponymer i de originale publikationer skal bevares, bortset fra rettelser, der er påbudt i ICN's Art. 60–61, hvilket forhindrer standardisering baseret direkte på efternavnet, som Forsskål selv skrev det. Efter at have diskuteret, hvad der skal rettes i henhold til ICN, og hvad der ikke skal, gennemgår vi de oprindelige eponymer og bestemmer deres korrekte ortografi, hvorefter antallet af varianter reduceres fra 34 til 16. Vi diskuterer også, om eponymerne er gyldige og legitime i henhold til ICN. Af de eponymer, der er baseret på Forsskåls efternavn, er syv ikke gyldigt publiceret, 30 er illegitime, og 36 er legitime (hvoraf en blev afvist af XVIII International Botanical Congress og en i øjeblikket foreslås afvist). Af de legitime eponymer er 19 epiteter for aktuelt accepterede navne på taxa.

Nøgleord: Egypten; *Flora Aegyptiaco-Arabica*; homonymer; fejlidentifikationer; ortografiske varianter; overflødige navne; Yemen.

Ib Friis
Professor emeritus
Botany
Natural History Museum of Denmark
University of Copenhagen
Universitetsparken 15
DK-2100 Copenhagen Ø
Denmark
ibf@snm.ku.dk
Orcid: 0000–0002–2438–1528

Mats Thulin
Professor emeritus
Systematic Biology
Department of Organismal Biology
Evolutionary Biology Centre (EBC)
Uppsala University
Norbyvägen 18D
SE-752 36 Uppsala
Sweden
mats.thulin@ebc.uu.se
Orcid: 0000–0001–5970–8343

Scientific plant names derived from the surname of Pehr Forsskål

– their history, orthography, legitimacy, and current use

by Ib Friis and Mats Thulin

Scientia Danica · Series B · Biologica · vol. 12

DET KONGELIGE DANSKE VIDENSKABERNES SELSKAB

Printed in Denmark by Narayana Press
Graphic design by Dan Eggers
ISSN 1904-5484 · ISBN 978-87-7304-467-4

Submitted to the Academy August 2023
Published April 2025

Table of Contents

1
Introduction

Pehr Forsskål (1732–1763) participated as a naturalist in the Royal Danish Expedition to Arabia Felix (1761–1767), the classical name for the south-western part of the Arabian Peninsula, now called Yemen. He died there, in the small town of Yerim [Yarim] in the western highlands, ca. 120 km south of the capital, Sanaa. The expedition took place in the early days of scholarly, non-dogmatic study of the Bible, and the overarching purpose of the expedition was to study local Arabic dialects, whether they could shed light on biblical philology, ancient features of Semitic culture, and the relationship between people and their environment. All this should be seen as the background for a better understanding of details in the Bible. One of the tasks of the naturalist was therefore to identify and study the flora and fauna, describe the organisms that were not previously described and given scientific names, and record their local names and uses, both covering the domesticated and the wild species.

On the initial part of the expedition, the participants travelled by ship from Denmark via the Atlantic and the Mediterranean to Cairo. Then the expedition had a long stay in Egypt and continued by ship from Suez along the eastern shores of the Red Sea. Whenever travelling by sea, Forsskål studied marine organisms. On land he studied the terrestrial flora and fauna, first near Marseille, then on Malta, at Constantinople [Istanbul], on the Aegean Islands, in lower Egypt, around ports of the Red Sea, and finally in many parts of Yemen, where he died in July 1763.

The first consignments of his more than 1800 collections of vascular plants, plus a small number of marine algae, were sent by himself to Copenhagen from Constantinople and Cairo, and until the departure from Egypt he continued sending seeds of plants to botanical gardens in Europe. His collections from the Red Sea, from its shores and from Yemen, as well as all his notes, were sent to Copenhagen from India by Carsten Niebuhr, the geographer and astronomer and only surviving member of the expedition.

A total of 775 new names of genera and species of vascular plants were proposed by Forsskål in his notes from the field. His notes on plants and animals were published posthumously by Niebuhr in the books *Flora Aegyptiaco-Arabica* (Forsskål 1775a), *Descriptiones animalium* (Forsskål 1775b), and *Icones Rerum Naturalium* (Forsskål 1776) with the help of an anonymous editor. From Marseille Forsskål recorded 265 species of plants, of which 13 were considered new, from Malta 87 species, of which 3 new, from around Constantinople and the Aegean Islands 481 species, of which 73 new, from Egypt 576 species, of which 206 new, and from ports along the Red Sea and lowlands and mountains in Yemen 693 species, of which 362 new.

This overview of Forsskål's botanical contributions to the results of the Royal Danish Expedition to Arabia Felix is here summarised from Christensen (1918) and Hepper & Friis (1994), and the study by Provençal (2010) of Forsskål's Arabian plant names. General historical and biographical details and broad evaluations of Forsskål's background, personality, biography and scholarly results have been published by Baack (2013a, 2013b, 2014) in a separate article and in two chapters dedicated to Forsskål, as well a general evaluation of the entire expedition in the book *Undying Curiosity. Carsten Niebuhr and the Royal Danish Expedition to Arabia (1761–1767)*.

The importance of Forsskål's botanical work on the expedition was first documented by the Norwegian-Danish botanist Martin Vahl, who in his works *Symbolae botanicae* (Vahl 1790, 1791, 1794) and *Enumeratio plantarum* (Vahl 1804, 1805) reidentified many of Forsskål's descriptions and specimens. But for more than hundred years, only very few European botanists were able to verify and supplement Forsskål's botanical studies by travelling in Yemen. When, in the second half of the 19th Century, travel in Yemen again became possible, the German botanist Georg Schweinfurth (1889, 1891) went to Aden and Yemen in 1888–1889, spent 80 days in the parts of the country that 127 years before had been visited by Forsskål, and collected during that journey 930 species of vascular plants, 273 more species than Forsskål had collected. Schweinfurth wrote that his journey should be seen as an act of respect, dedicated to the work of Forsskål, and he therefore distributed the duplicates of his collections from that expedition under the caption on the labels "IN MEMORIAM DIVI FORSKALII in Arabia Felici atullit G. Schweinfurth", using in capital Roman one of the Latinized variants of Forsskål's name.

Forsskål's plant names were used in the overview of Forsskål's works on plants and botanical collections by Hepper & Friis (1994), as well as in currently used works on the flora and vegetation of Egypt and Yemen and similar works on many surrounding countries. A taxonomical and nomenclatural revision of the plants in Forsskål's *Flora Aegyptiaco-Arabica*, as well as of the Forsskål specimens from the expedition to Egypt and Arabia, was published by Hepper & Friis (1994).

In the introductory parts of the present work, we have discussed the general rules guiding Forsskål's plant names, including eponyms, and the complications we see with regard to orthography and legitimacy. In the systematic part of this work, we have analysed and discussed all the eponyms of plants based on Forsskål's name, evaluated those names as cited by Hepper & Friis (1994), and made updates and corrections wherever necessary.

There is a need of correct orthography of Forsskål's eponyms to support their use in botany, particularly in nomenclatural databases in order to allow these databases to interact, and this work was prepared at the request of staff members at *The International Plant Names Index* (IPNI continuously updated). Databases provide a quick opportunity to check names of vascular plants, their spelling, nomenclatural and taxonomic status and geographical distribution, including the databases IPNI (continuously updated), *Global Plants* (continuously updated), *Plants of the World Online* (POWO continuously updated), *World Flora Online* (WFO continuously updated), *Global Biodiversity Information Facility* (GBIF continuously updated), and *The Taxonomic Name Resolution Service* (TNRS continuously updated). These originated initially all from the printed *Index Kewensis* (Jackson 1895a, 1895b) and older nomenclatural indices. The early indices and *Index Kewensis* have recorded Forsskål's eponyms, and it has sometimes in this work been necessary to discuss their interpretation of Forsskål's eponyms. Govaerts *et al.* (2021) described the development of the Kew databases, including IPNI and POWO, and the relations between these and other vascular plant databases.

Currently, the different databases do not agree in all aspects. It is hoped that this account, for which original references and protologues have been consulted for all names, will be a contribution towards future standardisation of Forsskål's eponyms. A note in IPNI (continuously updated) was added about Forsskål's eponyms: "In IPNI we have refrained from changing related epithets [eponyms based on Forsskål's surname] (other than transcribing diacritical signs), pending a more comprehensive review in the near future." We hope that the analyses in this work will bring us closer towards this goal.

In the systematic part of this work, we provide an overview of Forsskål eponyms, cite the names as originally published, the orthographic variant which we find is best in agreement with the ICN as we interpret it (see *Correcting or not correcting* ...). We have also for each name discussed whether it is validly published or not, or legitimate or illegitimate under the ICN and cited what appears to us to be the name in current use for the genus (only the generic name *Forsskaolea*) or the species.

Eponyms in botanical nomenclature

Eponyms are common in botanical and zoological nomenclature. Well-known botanists and zoologists, both collectors in territories that were botanically little known and scientist working in herbaria and museums, may sometimes have hundreds of species named after them, with the eponyms based on their surnames. Searches in the database IPNI (continuously updated) suggest that Nathaniel Wallich (1786–1854), field botanist in Nepal, India and Indochina, since ca. 1810 associated with the Calcutta Botanic Garden and in 1828–1832 working in London to name and sort the herbarium of the East India Company, has been awarded with more than 800 eponyms, which seems to be one of the highest numbers of eponyms awarded to a single person. The two Kew botanists, William Jackson Hooker (1785–1865) and Joseph Dalton Hooker (1817–1911), father and son, have together been awarded more than 1200 eponyms. For many other travellers and scientists, the number of eponyms might be in the few hundreds or below, but great 19th century travelling naturalists, such as Friedrich Welwitsch (1806–1872), who collected in south-western Africa, and Richard Spruce (1817–1893), who collected in tropical South America, each reach a number between 400 and 600. Carl Peter Thunberg (1743–1828), who travelled in South Africa, Indonesia and Japan and afterwards worked as a taxonomist in Uppsala, reaches over 400 (data from IPNI continuously updated).

In *Philosophia Botanica* (Linnaeus 1751: 171, aphorism 238), Linnaeus stated about generic eponyms and their justification that "Nomina generica, ad Botanici optime meriti memoriam conservandam constructa, sancte servanda sunt." [Generic names formed to preserve the memory of a well-deserving botanist; they should be meticulously observed]. Candolle (1813: 246), after several pages of recommendations about the formation of plant names of species based on morphological characters of the named plants or based on the place of origin, wrote slightly less enthusiastically about eponyms: "On peut, enfin, donner aux espèces le nom du botaniste qui les a décrites le premier, qui les à découvertes, ou qui a contribué à les faire connaître; ..." [Finally, one can name species after the botanist who first described them, discovered them or helped to make them known; ...].

These rules were almost unaltered incorporated in the first set of rules of botanical nomenclature, *Lois de la nomenclature botanique* (Candolle 1867: 22, Art. 32), which stated, again mentioning eponyms as species names last: "Le nom spécifique ... [s]'il est tiré d'un nom d'homme c'est ordinairement pour rappeler le nom de celui qui l'a découverte ou décrite, ou qui s'en est occupé d'une manière quelconque." [The species name ... if it is derived from a person's name, is usually given to recall the name of the person who discovered or described it, or who was occupied with it in some way].

This tradition has largely been followed until fairly recently, but sometimes also sponsors or politically powerful people associated with the establishment of colonies were remembered, and now a steadily wider circle of people are being honoured by an eponym of a plant or an animal. This has recently led to a discussion about the relevance and political correctness of eponyms, both new and old. In particular, the justification of eponyms has been debated in relation to honouring people completely unrelated to botany or zoology, or in a quite heated debate about eponyms remembering people who have acted in ways that are currently considered unethical. In this debate, Guedes *et al.* (2023) made the most radical claim about the continued use of eponyms, advocating ultimately to rename all species, and presumably also all genera with eponyms to better connect with local geography and culture. Guedes *et al.* conclude that "in the long term, ... removing all valid eponyms from biological nomenclature is the most ethical option."

The present work pays attention to all organisms given Forsskål eponyms, although we deal in depth only with plant names. Forsskål's eponyms were and are not given for the reasons reviewed, discussed, and rejected by Guedes *et al.* (2023), but rather for the reasons cited above from Linnaeus (1751), Candolle (1813) and Candolle (1867), that is to honour Forsskål's scientific work with animals and plants with due attention to the knowledge gathered and preserved by the local communities. Forsskål does not stand out with a particularly high number of eponyms, which is not surprising. He collected in parts of the world where few had made scientific studies in the Linnaean tradition before, or had collected plants or animals for permanent documentation in herbaria or museums according to the European scientific practise. Therefore, nearly all Forsskål's new species were described and named by himself in his manuscripts and on his collections, and published in his own posthumous works. Most of the important 18th century collectors with many eponyms did not publish all their material themselves, and their eponyms were awarded by scientists working in the herbaria or museums when they discovered that a particular genus or species was new to science.

The main reason why Forsskål was awarded eponyms was taxonomic changes or errors that later authors thought would require correction and then often called for an eponym remembering the corrected botanist. Forsskål's eponyms stand out as being either corrections of the wrong identifications he had made in the field with a limited amount of literature at hand or being later authors' misunderstandings of his posthumously published works. Since these misunderstandings to some extent still persist today, and the orthographic variation of Forsskål's eponyms is so striking, we have found it relevant to give a general overview.

Typically, all eponyms honouring the same scientist are spelt consistently by their different authors and are validly published. The eponyms commemorating W.J. Hooker and J.D. Hooker include more than 800 epithets as a substantive in genitive, *hookeri,* the rest as adjectives, over 300 as *hookeriana*, nearly 90 as *hookerianum*, and more than 50 as *hookerianus* (and very few other variants) and are mostly validly published and legitimate according to the ICN. However, the situation is far more complicated with regard to Forsskål's eponyms, not least because of the confusion about how his surname should be spelt, the presence of the special Scandinavian letter 'å' in his name, and the confusion caused by the difficulty for later botanists to understand the posthumously edited and published *Flora Aegyptiaco-Arabica*. In the following, we will outline these three complications.

The spelling of Forsskål's surname

While there is little doubt about the spelling of the surnames of most botanists with many eponyms, there has, as discussed by Friis & Thulin (1984), been much uncertainty about the spelling of Forsskål's surname, and also on its use in eponyms. The findings by Friis & Thulin (1984) are therefore expanded upon here to provide further evidence for this long-delayed follow-up on our note from 1984.

As already pointed out by Fries (1912: 110) and in the *Swedish Biographical Dictionary* (Anonymous 1964–1966), surnames were variously spelt by various members of Forsskål's family, and by various of their correspondents: Forskål, Forsskål, Forskåhl, Forskol, Forschkål, etc. However, Forsskål himself, in all his known correspondence in Swedish (Fries 1912; Christensen 1918; and in the preserved correspondence with Linnaeus in the Linnean Society of London, https://linnean-online.org/view/correspondence_by_collection/correspondence/Forssk=E5l=3APehr=3A=3A/Linnaeus=3ACarl=3A=3A.html, see Fig. 1), used the spelling with 'å' and double 's', while in some of his correspondence in German, including correspondence with Danish authorities in Copenhagen and the Danish envoy to Istanbul, he used the spelling 'Forsskaal' (see Christensen 1918, where a signature in Forsskål's hand and with a spelling as 'Forsskaal' is reproduced below the portrait opposite the title page).

Fries (1912) has stated that he did not recall having seen Forsskål himself use any other variant than 'Forsskål', but Fries was not aware of the German letters quoted by Christensen (1918). Hildebrand & Matinolli (1964–1966) seemed also to have been unaware of a variant spelling with 'aa'. Forsskål's contemporaries used various spellings in correspondence. For example J.P. Falck, a pupil of Linnaeus, used in his correspondence with Linnaeus alternatively either

'Forsskål' or 'Forsskåhl' (Fries 1912: 44). Fries (1912: 110) also mentioned that Forsskål's father, Johannes Forsskåhl (1691–1762) used a spelling with 'åh', and according to biographical notes on the Forsskål-family (Forsskåhl 1927) his brothers: Jonas Forsskåhl (1725–1783) and Johan Christian Forsskåhl (1726–1756) also used that spelling. The latter brother, a clergyman in Helsingfors interested in botany, corresponded with Linnaeus using the variant Forsskåhl (Fries 1912: 108), also used by more remote members of the family (Anonymous 1964–1966). However, in the article about Forsskål's father in *Swedish Biographical Dictionary*, Hildebrand (1964–1966) used the form 'Forsskål' rather than 'Forsskåhl'.

It is common that variants of Swedish personal names have an added 'h' after any vowel, often 'å' or 'ö', or even after a consonant. A few examples: The surname of the Swedish entomologist Carl Stål, identical with a Swedish word meaning 'steel', has when used by others a common variant with an 'h', Ståhl. The surname Öhrn is derived by adding an 'h' to a Swedish word 'örn', meaning eagle. The surname Svahn is similarly derived from a Swedish word 'svan', meaning swan, and Ahlm is derived from a Swedish word 'alm', meaning elm, etc. This is part of a tradition called 'decorative spelling' of surnames (Noreen 1913), a tradition also followed by the Forsskå[h]l family, but not by Forsskål, the naturalist of the Arabian Journey.

We have not seen the original letters from Forsskål to his professor in Göttingen, J.D. Michaelis, the initiator of the expedition to Egypt and Yemen, but in all letters in the published correspondence (Buhle 1794: 339–418, 489–492; 1795: 31–38, 56–66, 195–204), Forsskål's name is spelt with one 's' and 'å'. We cannot be sure if the consistent use of one 's' and 'å' represented spelling in the original letters or was introduced by the editor when preparing the correspondence for publication. Buhle (1794: XVIII) wrote in his introduction to the letters: "[Ich habe] die Orthographie zu Verbessern mir die Freyheit genommen." [I have taken the liberty to improve the orthography].

Linnaeus used the spelling 'Forsskål' in scholarly publications at least twice. In 1758, when Forsskål had returned to Uppsala from Göttingen, Linnaeus named a European moth after Forsskål with the spelling *Phalaena (Tortrix) Forsskåleana* L. (Linnaeus 1758: 531), now according to the Zoological Code with suppression of 'å' to 'a', as *Croesia forsskaleana* (Linnaeus 1758) or *Acleris forsskaleana* (Linnaeus 1758). Linnaeus used the same spelling in his commemorative note on Forsskål (Linnaeus 1764), in fact a brief obituary, when he had come to know about Forsskål's death in Yemen, and he named a new genus of the nettle family (Urticaceae), spelt *Forsskålea*, in Forsskål's honour, based on material grown from seeds sent by Forsskål from Egypt. The name of the genus is now accepted as *Forsskaolea* L.

Forsskål's well known, but politically controversial pamphlet on citizens' right to free speech was published without the author's name (Anonymous [Forsskål] 1759), but the extensive literature on Forsskål's importance for the development of ideas about civil liberty (in part reviewed in Carlsson & Goldberg 2017), seems consistently to use the spelling 'Forsskål'.

Although the posthumously edited *Flora Aegyptiaco-Arabica* (Forsskål 1775a), *Descriptiones animalium* (Forsskål 1775b), and *Icones rerum naturalium* (Forsskål 1776), have the variant spelling 'Forskål' on the title page, Forsskål's own variant is used almost consistently in the extensive biographical and nomenclatural literature (e.g. Christensen 1918, 1922; Schück 1923; Daniels & Stafleu 1972 [where 'Forsskål' is being used consistently in the biographical notes]; Wolff 1980; Rasmussen 1990; TL-2 2024). The spelling 'Forsskål' is also consistently used in the text by Hepper & Friis (1994), although the printers did not use an 'å' on p. iii, the title page, which demonstrates how difficult it is to obtain a completely consistent spelling in printed works. In his analyses of The Royal Danish Expedition to Arabia (1761–1767), Baack (2013a, 2013b, 2014) consistently used the spelling 'Forsskål', including in the biographical paragraph on Forsskål on pp. 55–63 and in the discussion of the botanical and zoological results of the expedition on pp. 331–336. We will therefore conclude that the correct spelling of Forsskål's name includes a double 's' and the letter 'å'.

Also Forsskål's Christian name has been very variably rendered, as Per, Pehr, Peter, Petter and Petrus, and Forsskål himself often wrote his Christian name differently in private and in formal letters. Forsskål's first Finnish biographer, Lagus (1877), and his Swedish biographer, Schück (1923), both preferred the colloquial form of the name, Petter. In his biography of Linnaeus, Fries (1903: 58–62) used Pehr, a shortened form of Peter, and Christensen (1918, 1922), in his detailed studies of Forsskål's travels and work, also used Pehr. In the first edition of Dansk Biografisk Lexicon [Danish Encyclopaedia of National Biography], Warming (1891: 258) used Per (and 'Forskål'). In the second edition of that biographical encyclopaedia, Christensen (1935: 183–185) used Petrus. In the third edition, Wolff (1980: 494) used Peter. The most commonly used names in this range are Pehr and Peter (TL-2 2024), where Pehr is cited as the preferred name. We have chosen to use Pehr, following Christensen (1918, 1922), [Daniels & Stafleu] (1972: 252), Friis & Thulin (1984), Hepper & Friis (1994), and TL-2 (2024). However, the spelling of Forsskål's Christian name is of no importance in botanical nomenclature.

The letter 'å' in Scandinavian languages and its transcription in plant names according to the international codes of botanical nomenclature

A particular point that caused considerable uncertainty outside Scandinavia in the spelling of Forsskål's name is the special Scandinavian letter 'å', and its history and significance in botanical nomenclature is therefore reviewed here. The sound represented by 'å' is pronounced approximately like the long vowel in the English word 'law', but it may also be a shorter and more open vowel. In mediaeval manuscripts from Sweden and Denmark, Skautrup (1953) has carefully traced the modifications of Roman and Gothic letters used to represent the sound of 'å' (see also Haugen 1976 and Sandersen 2002) and cited in particular a group of Swedish manuscripts that consistently use the letter 'å' (Skautrup 1953: 298–299, with figure), probably dated to be from before 1470. However, Skautrup refers to other 15th Century manuscripts where 'å' is used alternating with 'aa', which had been used in older texts for the same sound. In Danish, the use of letters in the 15th Century manuscripts is variable, but 'å' was very rarely used, and Danish scribes usually represented the sound 'å' with 'aa' (Skautrup 1953: 292–295; Sandersen 2002).

Skautrup (1953: 292) points out that the use of diacritical signs over vowels to modify their pronunciation was very frequent in 15th Century Low German manuscripts, mostly represented by a small 'e' over 'a', 'o', and 'u', and that this use later developed into the *Umlaut* still used in German and Swedish. However, he only traced two scribes from late Mediaeval Lübeck, the North German town with closest connections with Scandinavia, that used a ring over an 'a' as a different diacritic sign from the common 'a' with a small 'e' (Skautrup 1953: 296).

The consistent and well-documented use of the letter 'å' in Swedish began not in manuscripts, but with printed religious books in Swedish, published at and after the Protestant Reformation. In 1525, the printer Jürgen Richolff the Younger [der Jüngere], was called to Sweden from Lübeck (Skautrup 1953: 299; Lohmeier 1994). Shortly after his arrival in Sweden, he printed a Swedish Book of Hours in which 'å' was not used, but soon after, his printing workshop moved to Stockholm, and in 1526 he printed two religious books with consistent use of 'å' (*En nyttwgh wnderwijsning* [A useful education, a religious textbook] and *Thet Nyia Testamentit på Swensko* [a Swedish translation of the New Testament]), both authored by the Swedish theologian Olaus Petri. In these books, the later Swedish letters 'ä' and 'ö' were printed as an 'a with a small e above' and 'o with a small e above', as done by contemporary North German printers. Probably most important for the solid and permanent establishment of 'å' in Swedish was the consistent use of the letter in Gustav Vasa's monumental Reformation Bible in

Swedish (Anonymous 1541), also printed in Uppsala by Jürgen Richolff the Younger.

Therefore, already in 16th century Swedish, 'å' was a commonly used letter, and the use of 'aa' for 'å' became quite outdated in Swedish from ca. 1550 (Skautrup 1953: 300). Written Norwegian, very close to written Danish, maintained the use of 'aa' for 'å' up to a linguistic reform in 1917 peculiar to Norway, when the use of 'å' for the sound previously rendered as 'aa' became correct. In Danish, 'å' was used in a few printed works in the 16th Century, but several proposals to introduce 'å' for 'aa' in printed Danish in the 19th Century were unsuccessful, although they were advocated by a number of internationally known Danish linguists. In fact, a new letter, a ligature of 'A+O' or 'a+o', written as 'Ꜵ' or 'ꜵ', was informally proposed for 'å' (Skautrup 1953: 309), presumably in order to avoid actually copying the Swedish 'å', or to use a letter resembling the Danish letters 'æ'. Finally, the Swedish 'å' was introduced in Danish by an official orthographic reform of Danish in 1948 (Skautrup 1953: 302–309; Haugen 1976). It may be worth noting that the ligatures 'Ꜵ' and 'ꜵ' are now occasionally used internationally in print and is represented in Unicode as 'U+A734' and 'U+A735', while 'Å' and 'å' are represented in Unicode as 'U+00C5' and 'U+00E5'.

Rules in botanical nomenclature to transcribe the German and Swedish 'ä' to 'ae', 'ö' to 'oe', and 'ü' to 'ue' have been in force since the first *Lois de la Nomenclature Botanique* was adopted at the first (unnumbered) International Botanical Congress in Paris 1867, where it appeared in Art. 27 (Candolle 1867). The wording about these transcriptions was unaltered in the Vienna Code (Briquet 1906) but had changed to appear in a Recommendation IV(c), where it also appeared in the Bruxelles Code (Briquet *et al.* 1912). The numbering of this Recommendation changed to XL(d) in the abbreviated Code from the Cambridge Congress (Rendle 1934), and again to XXXIX in the full Cambridge Rules (Harms *et al.* 1935) and in the Brittonia-edition of that Code (Camp *et al.* 1947). The wording and place in Recommendation XXXIX was upheld in the Amsterdam Code (Sprague *et al.* 1950), but the number of the Recommendation changed to 82B in the Stockholm Code (Lanjouw *et al.* 1952).

The rule to suppress the use of the letter 'å' in plant names or to transcribe it to 'ao' was first introduced at the VIII International Botanical Congress in Paris in 1954 and was incorporated as part of Art. 73 in the Paris Code (Lanjouw *et al.* 1956). There was no proposal to transcribe 'å' to 'ao' in the published *Recueil Synoptique des Propositions* for the nomenclatural session at Paris (Lanjouw 1954), where the relevant proposals regarding orthography were dealt with under Art. 82. In Art. 82, Proposal A (J.P.M. Brenan in *British Botanists* 1954: 29 proposed: "The use of letters foreign to botanical Latin or of diacritic signs is to be regarded as an orthographic error, and such letters are to be transcribed and the signs suppressed.") Janchen (1954: 39) argued in Recommendation 82B Proposal B also for transcription of all diacritical signs: "[Man sollte] die betreffenden Buchstaben in angemessener Weise transkribieren." [One should transcribe the relevant letters in a pleasant way]. Neither Brenan, nor Janchen mentioned 'å'. At a pre-congress meeting in Geneva, it had been decided to establish a Committee for Orthography to report to the nomenclature session at the Paris Congress. The report by M. Pichon, H.W. Rickett and T.A. Sprague, *Rapport du Comité Spécial pour les Questions d'Orthographe* is included in Rousseau *et al.* (1955: 167–172). Brenan's proposal would, in the opinion of the committee,

> [make it mandatory] to eliminate accents, umlauts, etc. from older names in which they may be found and would prohibit their use in the future. Furthermore, these provisions for transcribing signs and letters not found in classical Latin would then concern all names, not only those derived from personal names. ... Several persons have called attention to the fact that to 'suppress' a diacritic sign may lead to the transformation of (e.g.) 'mülleri' to 'mulleri', which would be an incorrect orthography.

The following wording [was] suggested by Rickett and supported by Sprague [only the paragraph about letters foreign to Latin and diacritic signs is cited here]:

> Diacritic signs are not used in botanical Latin. In names (either new or old) drawn from words in which

> such signs appear, the signs must be suppressed with the necessary transcription of the letters so modified; for example, ä, ö, ü become respectively ae, oe, ue; é, è, ê become e, or sometimes ae; ø becomes oe. … It is suggested that the exact wording may be left to the Editorial Committee, if the proposed change is accepted in principle. It may be pointed out that the more sweeping disposal of diacritic signs in Pichon's version ('all [diacritic signs] to be suppressed') may perhaps lead to trouble.

Pichon's full proposal is not cited here. The report of the special committee on orthography was unanimously accepted as a whole packet by the Paris nomenclature session, and the Editorial Committee was charged to redraft the Paris Code in accordance with this report and make the necessary corrections (Rousseau *et al.* 1955: 166).

We have found no printed documentation of how, after the Paris Congress, the transcription of 'å' to 'ao' in the renumbered Art. 73 became included in the Code without having been recommended by the *Comité Spécial pour les Questions d'Orthographe*. The committee's recommended transcription of 'ø' to 'oe' was included in the French and German texts of the Paris Code (Lanjouw *et al.* 1956: 90, 140), but was left out, presumably by error, in the English text (Lanjouw *et al.* 1956: 46). In the eleven codes (and in 2025 also the coming Madrid Code) following the Paris Code, the rule about transcription of 'ä', 'ö' and 'ü' has been perpetuated, together with the new rule on 'å', and from the Montreal Code (Lanjouw *et al.* 1961: 53), the rule for transcription of the Danish and Norwegian 'ø' to 'oe' was also included in the English text of Art. 73.

From the Berlin Code (Greuter *et al.* 1988), the text of the Code was published in an English version only, and a ruling was added to Art. 73.5 that "the ligatures -æ- and -œ-, indicating that the letters are to be pronounced together (*Arisæma, Schœnus*), are permissible." The ligature '-æ-' has been much used in botanical texts and scientific plant names, but in most typography the ligature is indistinguishable from the Danish and Norwegian 'æ', which indeed was originally also a ligature. In the Tokyo Code (Greuter *et al.* 1994) all the contents about orthography from Art. 73 was transferred to Art. 60, and the total number of articles in the Division II of the Code was reduced to 62. In the new Art. 60.7, the main part about diacritic signs was transferred unaltered from the Berlin Code, but the wording about the ligatures 'æ' and 'œ' was changed to "the ligatures -æ- and -œ-, indicating that the letters are to be pronounced together, are to be replaced by the separate letters -ae- and -oe-."

Stearn (1973) has, as part of his now classical textbook *Botanical Latin* written a guide to the formation of genus names and epithets in Latin, including a historical review, and a section on epithets commemorating persons. However, in the case of permissibly Latinizing the name 'Forsskål', Stearn's work offers no help, only referring to the Montreal Code and stating that the syllables in the epithet not modified by the rules for the endings retain their original spelling, unless they contain letters foreign to Latin plant names or diacritic signs. Stearn also quoted that 'å' should become 'ao'.

In spite of the limited changes since the Paris Botanical Congress in 1954 with regard to the Scandinavian letters 'ä', 'æ', 'ö', 'ø', and 'å', these letters have been continued to be discussed. Before the XI Botanical Congress in Seattle in 1969, Paclt (1968) proposed that 'å' should be transliterated in plant names as 'aa' for the sound shared by 'å' and 'aa' in agreement with the old Danish and Norwegian tradition, rather than to 'ao'. Paclt's argued, as an example: "The name of *Forskohlea* (= *Forskalea*, *Forskaolea*, after Forskal), a Linnean genus of the Urticaceae, must not be altered to *Forskaalea* since å did not appear in the original spelling." Paclt obviously accepted the publication of *Forsskaolea* from the reprint in the *Amoenitates academicae,* vol. 7, ed. 1, 1769, and Vol. 7, ed. 2, 1789. Hylander (1968) objected that the original publication was actually a thesis from 1767 and pointed out that in that work, Linnaeus consistently used the spelling *Forsskålea* which, under Paclt's proposal, would become *Forsskaalea*, "one of the few spellings which do not seem to have been used hitherto of this so often

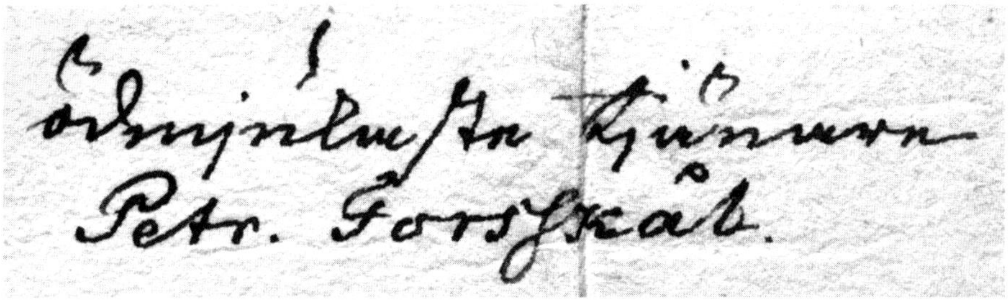

Fig. 1. Forsskål's signature on a letter to Linnaeus, written 10 Dec. 1754 and sent from Göttingen. The letter, L1685, is the oldest of the 28 letters to Linnaeus from Forsskål, all now preserved at the Linnean Society of London (transcribed by Fries 1912 as no. 1337). From this and until his death, Forsskål always wrote 'ss' in his name, consisting of first a small 's', followed by a long 's' that resembles a handwritten 'f', a 'k' like a printed letter, then the letter 'å', and finally the letter 'l'. His salutation to Linnaeus, written in Gothic handwriting, also remained: "högwälborne Herr Archiaterns och Riddarens / ödmjukaste tjänare / Petr. Forsskål." [humblest servant of the noble Mr. Royal Chief Physician and Knight, Petr. Forsskål]. Reproduced with permission of the Linnean Society of London.

mistreated name." Paclt's proposal was not accepted at the Seattle Congress or incorporated in the Seattle Code (Stafleu *et al.* 1972).

A similar view of how to deal with 'å' in biological nomenclature was offered by Regnéll (1952), palaeontologist at the University of Lund with specialisation in echinoderms. Regnéll stated that transcription of 'å' to 'ao' appears strange to Scandinavians, but he rejected to transcribe 'å' to 'a' because it would be identical to 'a' transcribed from 'ä'. Therefore, like Paclt, Regnéll recommended the adoption of 'aa' as the transcription for the Scandinavian 'å', arguing that this had been a long-lasting official use for the identical sound in Denmark and Norway. This would prevent homonymy between originally not identical names.

In a report to the XIII International Botanical Congress in Sydney, Australia, in 1981, Nicolson (1981: 178) summarised the findings of the *Special Committee on Romanization of authors' names from non-Roman scripts*. Although this report mainly deals with full or abbreviated authors' names and not specifically with eponyms as scientific plant names, the report contains some observations of relevance to the present subject. In the part of the report entitled 'ø, å = oe, aa' it is suggested that the pronunciation of the letter 'å' "approximates a long o ('Oh!' is written 'å!' in these [Scadinavian] languages). Actually it is somewhere between a long a (as aw in law) and a long o (as ow in low). The letter is very old." The committee concluded that

> the issues [about transcribing 'å' to Roman scipt] are confused by occasional past usages (in Latinization) of writing 'å' as 'ao', 'ah', 'oh', and sometimes simply 'a'. This led to an amazing series of orthographic variations [for example in the genus name based on Forsskål's name], such as *Forsskaolea*, *Forskahlea*, *Forskohlea*, *Forskalea*, etc. This disarray was solved by passing what appears in Art. 73.6, "In names … drawn from words in which such [diacritic] signs appear … å becomes ao …" While it may be desirable to leave ao for å when it appears in plant names (if only to leave *Forsskaolea* undisturbed), it is definitely undesirable that this be done for authors' names. … There is substantial agreement that, except in plant names, å should not become ao. Transliteration as aa can be considered [in authors' names], as well as simply a, but it would be preferable

to avoid the whole controversy by maintaining å if that is what the original author uses.

In fact, from data in the *Index Nominum Genericorum* project Stafleu & Cowan (1976: 856) had listed the orthographic variants of Linnaeus's original *Forsskålea* (Linnaeus 1764) as *Forshohlea* Batsch (1802); *Forskaela* Cothen. (1790); *Forskaelea* Scop. (1777); *Forskalea* Juss. (1789); *Forskaohlea* C.Agardh (1825); *Forskaohlia* Webb & Berthel. (1844–1850); *Forskoehlea* Reichard (1778); *Forskoelea* Brongn. (1843); *Forskohlea* L. (1767); *Forskolea* (1769); *Forskolia* Wight (1853). The authors of these names are not listed in *Literature cited* in this work, but their abbreviated names can be found under *Authors* in IPNI (continuously updated). A similar list for orthographic variants of epithets in secondary literature has not been produced, but it is almost certain that it would be much longer. In this work, we focus on original publications of epithets in basionyms, new combinations, and replacement names.

As a consequence of our above account of the history of 'å' in writing and in botanical nomenclature, we accept, in agreement with ICN, Art. 60.7, that when 'å' appears in the original publication of a Forsskål eponym, it is transcribed to 'ao'. More difficult to deal with are past occasional writing 'å' simply as 'a', or as 'ao', 'ah', or 'oh', when an 'å' is not used in the eponym in the original publication. In a chapter below, we cite an example of 'Å' being intentionally Latinized as 'A' in the original publication of an eponym: *Angstroemia* Bruch & W.P.Schimper (Latinized from Ångström), and accepted as an example of intentional Latinization in agreement with the Code by the XVI International Botanical Congress in St. Louis, USA, in 1999.

The publication of Forsskål's notes and later authors' difficulties with interpretation of *Flora Aegyptiaco-Arabica*

After Forsskål's death in Yemen, Niebuhr sailed from Mokka in Yemen to Bombay in India on a ship belonging to the British East India Company and was able to carry with him Forsskål's collections and manuscripts from the Red Sea and Yemen. No Danish ship called at Bombay, and Niebuhr therefore sent Forsskål's herbarium to the Danish trading post, Tranquebar, from where it eventually was sent to Copenhagen on Danish ships. Niebuhr was hesitant to send Forsskål's manuscripts and accompanying drawings by ship, and he thought of carrying them with him himself on his onwards journey.

It is not known if the manuscripts finally came to Copenhagen on a ship or over land with Niebuhr (Christensen 1918: 59). In 1767, when Niebuhr arrived at Copenhagen from India, he was obliged to publish the official travel reports of the expedition before he could deal with the publication of Forsskål's manuscripts and drawings (Niebuhr 1772, 1774–1778). These books were costly works to produce due to the many drawings that had to be published as engravings. Niebuhr paid for the printing of these books himself, although the Danish King contributed to the costs of the engravings (Pedersen 2004). Niebuhr is indicated as the editor in the introductions to both the *Flora Aegyptiaco-Arabica* and the *Descriptiones animalium*, and, due to the mentioning of personal memories from the expedition, it seems likely that indeed he provided text for the introductions. However, according to the biography written by Niebuhr's son, the politician and historian Barthold Georg Niebuhr (1776–1831) (Niebuhr 1816, 1817), his father later regretted that he had been attributed this editorship, because, not being acquainted with botany and zoology and not well versed in Latin, he had also paid out of his own pocket an unidentified Swedish naturalist to edit Forsskål's manuscripts from the field and compile these notes into publishable books, and the results of this editorial work was not everywhere well received (Niebuhr 1816, 1817; Pedersen 2004).

Christensen (1918) has pointed out that the early reception of Forsskål's *Flora Aegyptiaco-Arabica* was favourable. Due to Forsskål's many new observations and hitherto undescribed species, the contemporary botanists overlooked or excused the obvious inconsistency in the editing or even errors that soon became apparent. Later, the reviews became less laudatory, particularly of the way the book had been edited, and in particular Martin Vahl, in the introduction to the first volume of his *Symbolae* (Vahl 1790), wrote a sharp criticism of the way in which the work had been finalised for the press. With Forsskål's almost unbelievable diligence, one could have expected a much better work, had he been allowed to publish his observations and discoveries himself, and not this work based on his edited field notes, published without comparing them with the specimens he had collected, and which were preserved in the Forsskål herbarium in Copenhagen. To this came a range of practical and formal shortcomings of the editor's work.

Later scholars studied the many new plants collected or named by Forsskål, and they often found, correctly or incorrectly, that there was a need to name or rename them, sometimes because the names proposed in Forsskål's manuscripts had been used by others for other species in the time between Forsskål's work on the expedition and the publication of the *Flora Aegyptiaco-Arabica*, but also for a number of other reasons related to the arrangement of the book, where they often overlooked that binomials published in one part of the book were connected with descriptions in another part. When renaming Forsskål's species, the later authors often used new names based on Forsskål's name, eponyms, perhaps with intentional varying Latinization, perhaps with orthographic errors in rendering the name Forsskål, 'this so often mistreated name' (Hylander 1968). In the following, we briefly review the way Forsskål's notes were published and the main reasons why the book was so frequently misinterpreted by later scientists.

On the expedition, Forsskål had compiled floristic lists of the plants which he had recorded from Egypt and Yemen, as well as for a number of localities visited in the Mediterranean, he had made descriptions of most of the plants he had collected, particularly those he considered new, and he had written a detailed phytogeographical essay on the flora of lower Egypt and a preliminary one on the flora of Yemen (Christensen 1918: 66; Hepper & Friis 1994: 35–38). Forsskål's manuscripts brought home from the expedition consisted of scraps of paper with lists and descriptions of animals and plants. In the foreword to the *Descriptiones animalium*, Niebuhr wrote about Forsskål's manuscripts (Forsskål 1775b: 7–19) that the observations on each species were drafted on separate slips of paper because Forsskål would then, when the number of species increased, be able to rearrange them in their new and correct place like sheets in a herbarium, while, if he had entered all the notes in one volume, there would have been no possibility for rearrangement. Niebuhr was fully aware that with this kind of loose-paged and flexible manuscripts, pages might have disappeared, particularly as the zoological and botanical manuscripts consisted of more than 1800 scraps of paper. Niebuhr also here stated that he left Forsskål's loose-paper manuscripts in the hand of a man truly knowledgeable in natural history and then did not give them more thought for a long time, until his own books reporting on the expedition had been published. About the best way of publishing the manuscripts, he had asked the advice of learned friends, who unanimously recommended that the words of the author would be more valuable than any revision, and that the public would want to hear precisely what Forsskål had discovered.

The book dealing with plants, the *Flora Aegyptiaco-Arabica*, appeared in one volume (title page reproduced in Fig. 2) with two parts and an introduction that was signed and probably partly written by Niebuhr, who had also drawn and engraved a useful map of Forsskål's collecting localities in Yemen (Fig. 3).

The first part of the *Flora Aegyptiaco-Arabica*, which has no separate title page, is a sequence of four floristic accounts with lists of names of species and abbreviated information about collecting locality and habitat, but sometimes also with short descriptions. We will in the following refer to them collectively as the *Florulae*. They are: *Florula littoris Galliæ ad Estac prope Massiliam. Accedit florula insulæ Maltæ.* [Small flora of the French coast at Estac near Marseille, to which is added a small flora of the islands of Malta]. *Flora Constantinopolitana, littoris ad Dardanellos et insularum Tenedos, Imros, Rhodi.* [Flora of Constantinople, of the coast of the Dardanelles and the islands of Tenedos (Bozcaada in Turkish), Imros (Gökçeada in Turkish), and Rhodos]. *Flora aegyptiaco: sive catalogus plantarum systematicus Aegypti Inferioris: Alexandriæ, Rosettæ, Kahiræ, Sues.* [Flora of Egypt, that is a systematic catalogue of plants of Lower Egypt: Alexandria, Rosetta, Cairo, Suez]. *Flora Arabico-Yemen. Sive catalogus plantarum Arabiæ felicis systematicus.* [Flora of Arabia-Yemen, that is a systematic catalogue of the plants of *Arabia felix*]. Apart from a page reviewing the new genera from Yemen (Fig. 4), the floristic lists in this part are arranged according to the order in which the main localities of the expedition were visited and has pagination with a sequence of Roman numerals from I to CXXXVI. The systematic arrangement is that of the Linnaean sexual system. Two examples of pages from this part of the book, the list from the Arabian coast and Yemen, *Flora Arabico-Yemen* (p. LXXIX–CXXVI), are reproduced here as Fig. 5 and 6. They document names discussed later in this work, and a single asterisk refers to a description of a new species later in the book, while two asterisks refer to a description of a new genus later in the book.

The main second part of the book, the *Descriptiones plantarum florae Aegyptiaco-Arabica*, contains detailed descriptions of new genera and species of plants from Egypt, the Red Sea Coast and Yemen with all taxa arranged into one systematic sequence according to the Linnaean system, and it has pagination with a sequence of Arabic numerals from 1 to 219. Two examples of pages with names discussed later among the Forsskål eponyms are shown here as Fig. 7 and 8. In the *Descriptiones,* the species from all the localities of Egypt, the Red Sea Coast and Yemen are not only listed in one common taxonomic sequence according to the Linnaean sexual system, but also divided into eight *Centuriae*, or groups of hundred species. This completely artificial arrangement is superimposed on the Linnaean sequence.

Although the new species were mostly described in detail in the *Descriptiones,* more than 250 species were only described with short diagnoses or descriptions in the floristic lists in the first part of the book (Christensen 1918, 1922; Hepper & Friis 1994: 32–34). The *Flora Aegyptiaco-Arabica* was published without an index, and the need for the reader to orient him or herself in the book was felt to be so vital that Christensen (1922) compiled and published a first comprehensive index to the *Flora Aegyptiaco-Arabica* as part of his revision of Forsskål's herbarium in Copenhagen. In the introduction to the first volume of his *Symbolae botanicae*, Vahl (1790; as translated by Fox Maule 1984), who had compared every detail of the *Flora Aegyptiaco-Arabica* with the herbarium material from the expedition existent in Copenhagen, was critical about the posthumous edition of the *Flora Aegyptiaco-Arabica* but passed this mild comment on the work in print: "Even if several particulars in this posthumous work, which lacked the ultimate correction from its authors hand, could be criticized for inaccuracy, it is still worthwhile to pay considerable attention to most of the observations in [it]."

Forsskål's works, the herbarium material collected by him and the names he gave to his new discoveries have right up to now attracted much attention from botanists. One important reason for this is that most of Forsskål's new species have wide distributions in the Mediterranean, in the Middle East, and, for the tropical Arabian species from Yemen, far into tropical Africa. The *Flora Aegyptiaco-Arabica* is therefore an early source for many plant names from a much larger area than that actually visited by the expedition, being of interest far beyond the borders of Egypt and Yemen. Another reason for the lasting interest in Forsskål's work is the nomenclatural complications derived from the imperfect editing of the *Flora Aegyptiaco-Arabica*, which, as stated by Thulin (2016), "has given rise to numerous nomenclatural problems and controversies of various kinds and will no doubt continue to do so."

To indicate full references to the names and other information in the *Flora Aegyptiaco-Arabica* is complicated, as references, in order to be all-inclusive, must cite both the floristic part paginated with Roman numerals and refer to the descriptive part in the *Descriptiones* paginated with Arab numcrals. See an examples of the *Florulae* list from Yemen in Fig. 5 and of the *Descriptiones* in Fig. 7 and 8. In this work, we cite references to the *Florulae* with Roman numerals first, including the number of the taxon. After that, we cite references to the *Descriptiones*, with the Arab page number first, followed by the number of the *Centuria* in which the species can be found, and finally the number of the taxon in the *Centuria*. As mentioned above, in the *Florulae* of the *Flora Aegyptiaco-Arabica*, there is a system of references which indicate if there is a description of a new species (one asterisk) or a new genus (two asterisks) in the *Descriptiones*. If this had been consistently worked out by the editor and properly understood by 18th and 19th century authors, it might have guided them to the places where both complete binomials and the associated descriptions could be found, and they might have avoided many of the later misunderstandings.

Another complication is that Niebuhr's anonymous editor did not place all Forsskål's descriptions of new species in the *Descriptiones*, particularly if the diagnoses or descriptions were short from Forsskål's hand. Many such short descriptions were placed with their binomials in the *Florulae*, particularly those drafted before Forsskål reached Egypt and Yemen. An example from Forsskål's visit to the island of Imros in the Aegean Sea listed in *Flora Constantinopolitana* can be cited as a case of these short diagnoses or descriptions. *Hordeum imrinum* Forssk. (Forsskål 1775a: XIX, No. 52) is thus diagnosed: "No. 52. [Hordeum] e) *imrinum. – Imr.* [the island of Imros, now in Turkish named Gökçceada] *In collibus, parcius. Flosculo medio hermaphr. aristato; lateralibus masculis muticis; aristis longis; culmo pedali. Facies* H. murinis*; forte tamen nova species*." [In the hills, sparsely. The central flower bisexual, awned; the lateral ones male and without awn; the awns long, the culm a feet long. Has the appearance of *Hordeum murinum*; nevertheless a new species.] This protologue validates the name *Hordeum imrinum* Forssk. Based on modern studies of the preserved material (Hepper & Friis 1994: 270), Forsskål's name is now considered a synonym of a subspecies of *Hordeum murinum* subsp. *glaucum* (Steud.) Tzvelev, in *Novosti Sist. Vyssh. Rast.* 8: 67, 1971, the name which has priority at subspecific rank.

Another peculiarity in *Flora Aegyptiaco-Arabica* is that protologues of new genera in the *Centuriae* may sometimes have no specific epithets, but except for one case, the new genus *Simbuleta*, full species names can be found in the *Florulae* (Friis 1984). The fact that later authors overlooked the connection between the two parts of the book is probably the most important reasons for the creation of the 73 Forsskål eponyms under the botanical code. The later authors wanted to fill a nomenclatural gap that was not there.

In other cases, when a separate generic description was not provided for new genera with more than one new species, there were indeed nomenclatural gaps in the *Flora Aegyptiaco-Arabica*. In such cases, none of the new names are validly published according to the ICN, and later authors' validly published new names are justified. Three names of new genera from Yemen discussed in this work, *Catha* (Celastraceae), *Suaeda* (Chenopodiaceae), and *Saelanthus* (Vitaceae), were without validly published generic names in the *Flora Aegyptiaco-Arabica* (listed in the enumeration in Fig. 4 as new genera, but nowhere with generic descriptions). Two of these genus names were later validated (*Catha* Forssk. ex Scopoli, Introd. 228, 1777; *Suaeda* Forssk. ex J.F. Gmelin, *Onomat. Bot. Compl.* 8: 797, 1776, *nom. cons.*) and are in current use, while *Saelanthus* Forssk. ex Scopoli, *Introd.* 102, 1777, is now a synonym of *Cissus* L.

An additional complication with Forsskål's names was the fact that *Flora Aegyptiaco-Arabica*, mainly in the *Florulae*, occasionally contained names that were formed as phrase names and not given binary Linnaean names (see an example in this work under *Helichrysum forskaohlii* (J.F.Gmel.) Hilliard & B.L. Burtt). This was not a unique feature of Forsskål's flora; it could also be found in other works from the first decades of Linnaean binary nomenclature, and this meant that these works, some of them valuable as sources of early names, could have been subjectively rejected under the international codes of botanical nomenclature up to the Tokyo edition of the ICN (Greuter 1984; Friis *et al.* 1984; Jeffrey 1985; Friis & Jeffrey 1986; Friis 1992; Greuter *et al.* 1994). After the changes made to ICN for its Tokyo edition, the early post-Linnaean works that might cause complications due to inconsistent use of Linnaean nomenclature can be evaluated for suppression by the General Committee of Botanical Nomenclature as a source of binary names, and with the approval of the following International Botanical Congress placed on a list of *Opera utique oppressa* (ICN, Art. 34.1 and Appendix 1).

Sometimes, in the time between Forsskål's death in Yemen and the date of the publication of *Flora Aegyptiaco-Arabica,* the name for new species proposed by Forsskål in his as yet unpublished field notes were used by other authors for unrelated species. Phillip Miller (1691–1771) published the name *Solanum villosum* Mill., *Gard. Dict.* Ed. 8.: Species No. 2, 1768, after Forsskål's death in Yemen, but before the publication of *S. villosum* Forssk., *Fl. Aegypt.-Arab.*: CVII, No. 151; 47, *Cent.* II, No. 57, 1775. A similar situation occurred with *Erica verticillata* P.J.Bergius, *Descr. Pl. Cap.*: 99, 1767, published from South Africa, but before the delayed publication of Forsskål's name *E. verticillata* Forssk., *Fl. Aegypt.-Arab.*: XXV, No. 187; 210, *Cent.* VIII, No. 43, 1775.

A peculiarity in the *Flora Aegyptiaco-Arabica* that caused many later authors to rename Forsskål's new species, was that Forsskål had identified plants collected on the expedition with species described by Linnaeus from America or Asia. Forsskål's plants were in fact unrelated new species from Egypt and Yemen, but they were misidentified by Forsskål on the expedition because he used a small library, mainly works by Linnaeus, of which *Species plantarum*, probably the first edition, because the second was published in 1762, after the expedition had left Copenhagen, and his *Systema naturae,* probably the 10th edition, published in 1758–1759, were the most important. Taken on face value as published in the *Flora Aegyptiaco-Arabica,* these Linnaean names could be either (1) Forsskål's misidentifications, made while he was still in the field, or (2) later homonyms. Different rulings exist in the ICN for these two categories: (1) names for misidentifications should be dealt with in agreement with Recommendation 50D, and they have no formal nomenclatural status; (2) later homonyms for distinct species, which would be illegitimate and to be treated in agreement with Art. 53.1 and other articles about illegitimate names. Hepper & Friis (1994: 25–26) noted that Forsskål on the expedition wrote to Linnaeus that he

> expected to find American and Asian plants in Egypt and Arabia … He seems often to have accepted an identification if the plant matched a description in the Linnaean works he had with him, no matter where the species was [originally] described from. He used the botanical books of Linnaeus as a world flora. In a letter to Linnaeus from Yemen, he wrote: "Here I have found a lot of American, Indian, and new plants …"

Hepper & Friis (1994) listed a total of 188 cases where Forsskål has wrongly used earlier Linnaean names for what were actually new species. A complete search for and analysis of all cases where Forsskål has used earlier Linnaean names for what were actually new species should eventually be made, but here only the situations that have resulted in eponyms based on Forsskål's surname are discussed.

From the evidence available to them, Hepper & Friis (1994: 25–26) assumed that it was premature to state whether Forsskål's use of Linnaean names for American and Asian plants were misidentifications or later homonyms. However, for this publication, we have decided to change this conclusion. From his writings, including his above-mentioned letter to Linnaeus, it is clear that he really thought he was finding and identifying American and Asian plants in Arabia, and when he dealt with species that we now know do not occur both in Arabia and in America, in remote parts of Asia or elsewhere, from where the Linnaean plant was described and named, we are convinced that Forsskål's use of Linnaean names from places far from Egypt or Yemen are cases of misidentifications. According to ICN, Rec. 50D, one should indicate a published misidentification in citation with the abbreviation 'auct. non', but we have chosen to refer to them as '*sensu* Forssk., non L.' (or a few other authors). Later authors were right in giving new names in these cases.

FLORA
ÆGYPTIACO-ARABICA.
SIVE
DESCRIPTIONES
PLANTARUM,
QUAS
PER
ÆGYPTUM INFERIOREM
ET
ARABIAM FELICEM
DETEXIT, ILLUSTRAVIT
PETRUS FORSKÅL.
PROF. HAUN.

POST MORTEM AUCTORIS
EDIDIT
CARSTEN NIEBUHR.
ACCEDIT
TABULA ARABIÆ FELICIS GEOGRAPHICO-BOTANICA.

HAUNIÆ, 1775.
EX OFFICINA MÖLLERI, AULÆ TYPOGRAPHI.
PROSTAT APUD HEINECK ET FABER.

Fig. 2. Title page of *Flora Aegyptiaco-Arabica.* The title page states that the book was *post mortem auctoris edidit Carsten Niebuhr* [posthumously edited by Carsten Niebuhr]. As explained in the text, it was published with the aid of an unknown editor. It was published in *Hauniæ* [Copenhagen] in 1775. Niebuhr spelt Forsskål's name with one 's'. The map mentioned on the title page, *Tabula Arabiae Felicis geographico-botanica,* is reproduced here in Fig. 3.

Fig. 3. Carsten Niebuhr's map of Yemen (published as frontispiece in the *Flora Aegyptiaco-Arabica*), showing Forsskål's collecting localities with connecting thin, stippled lines that indicate Forsskål's travel route.

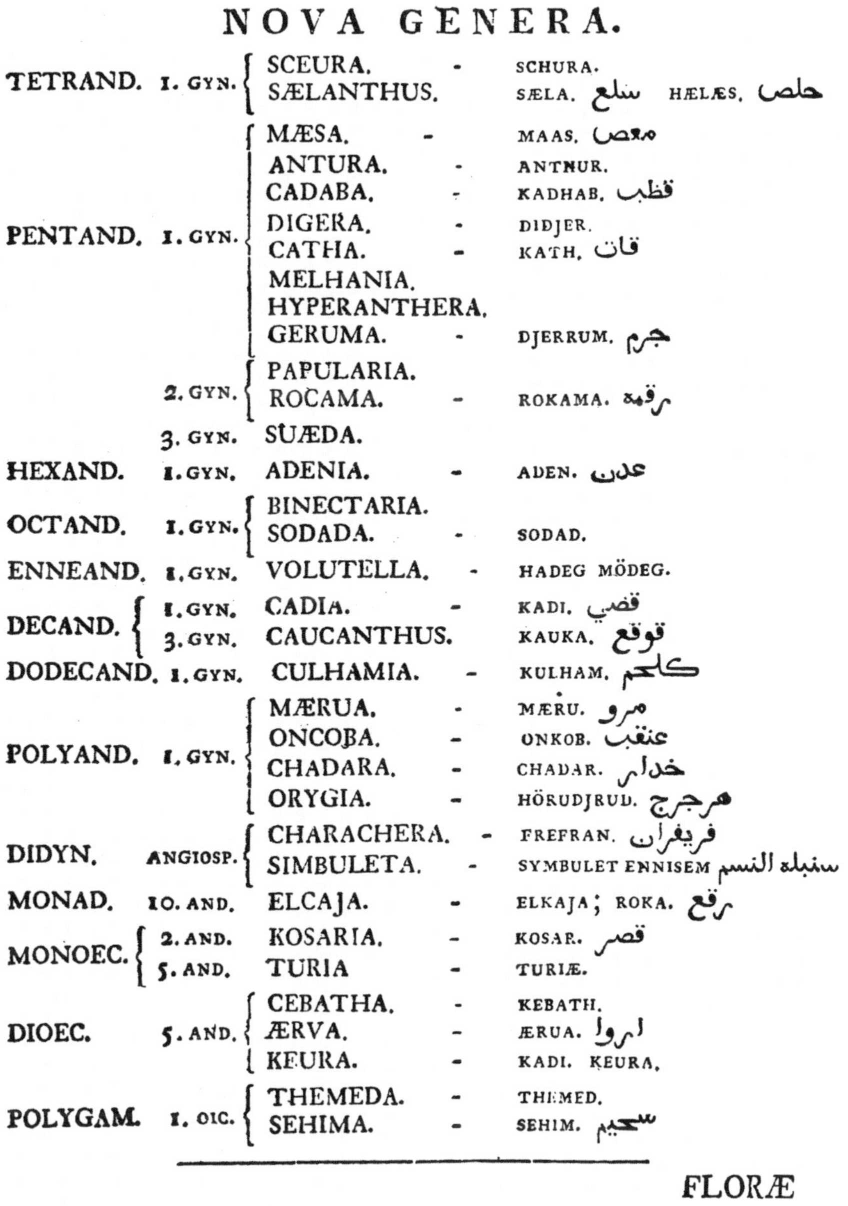

NOVA GENERA.

TETRAND.	1. GYN.	SCEURA. -	SCHURA.
		SÆLANTHUS.	SÆLA. سلع HÆLÆS. حلص
PENTAND.	1. GYN.	MÆSA. -	MAAS. معص
		ANTURA. -	ANTHUR.
		CADABA. -	KADHAB. قظب
		DIGERA. -	DIDJER.
		CATHA. -	KATH. قات
		MELHANIA.	
		HYPERANTHERA.	
		GERUMA. -	DJERRUM. جرم
	2. GYN.	PAPULARIA.	
		ROCAMA. -	ROKAMA. رقمه
	3. GYN.	SUÆDA.	
HEXAND.	1. GYN.	ADENIA. -	ADEN. عدن
OCTAND.	1. GYN.	BINECTARIA.	
		SODADA. -	SODAD.
ENNEAND.	1. GYN.	VOLUTELLA. -	HADEG MÖDEG.
DECAND.	1. GYN.	CADIA. -	KADI. قصي
	3. GYN.	CAUCANTHUS.	KAUKA. قوقع
DODECAND.	1. GYN.	CULHAMIA. -	KULHAM. كلحم
POLYAND.	1. GYN.	MÆRUA. -	MÆRU. مرو
		ONCOBA. -	ONKOB. عنقب
		CHADARA. -	CHADAR. خدار
		ORYGIA. -	HÖRUDJRUD. هرجرج
DIDYN.	ANGIOSP.	CHARACHERA. -	FREFRAN. فريفران
		SIMBULETA. -	SYMBULET ENNISEM سنبلة النسم
MONAD.	10. AND.	ELCAJA. -	ELKAJA; ROKA. رقع
MONOEC.	2. AND.	KOSARIA. -	KOSAR. قصر
	5. AND.	TURIA -	TURIÆ.
DIOEC.	5. AND.	CEBATHA. -	KEBATH.
		ÆRVA. -	ÆRUA. ارول
		KEURA. -	KADI. KEURA.
POLYGAM.	1. OIC.	THEMEDA. -	THEMED.
		SEHIMA. -	SEHIM. سحيم

FLORÆ

Fig. 4. Page LXXX in the floristic part of the *Flora Aegyptiaco-Arabica*, listing the new genera mainly described in *Flora Arabico-Yemen*. This page indicates the systematic position of the new genera in the Linnaean system, the name of the genus and its vernacular Arabic name in Latin and Arabic characters. Of the new genera discussed here, *Sælanthus* has two different vernacular names, *sæla* and *hælæs*, indicated with Latin and Arabic characters, while *Digera, Kosaria, Catha*, *Caucanthus, Turia* and *Themeda* have vernacular names of one word, respectively. The Arabic names are recognisable in Forsskål's Latinized generic names: *didjer* [*Digera*], *kosar* [*Kosaria*], *kath* [*Catha*], *kauka* [*Caucanthus*], *turia* [*Turia*], and *themed* [*Themeda*]. For *Simbuleta*, Forsskål has recorded a vernacular name of two words, *symbulet ennisem*, but only Latinized one for the new genus name [*Simbuleta*]. According to Provençal (2010: 92) the full Arabic name means 'the corn ear of the breeze.'

CXV

367. (*) OCYMUM *f) badienſe.* Mia. *Medân.* Öude. Blg.
368. (*) Novum genus? Zatarhendi. *g) ægyptiac.* CP. *Medân.* مدان Btf. (Ægyptis Zatarhendi.)
369. (*) - *h) villoſum.* Ps. Taæs.
370. (*) - *i) vaalæ.* CP. Btf. *Vaalæ.* واله
371. SCUTELLARIA *galeric.* - Mm. Ms. - Kurma.] Foliis ovato-lanceolatis, crenatis.
372. PHRYMA? - - - - - *Hamſched.* Hadîe. Muſa. Seminibus 2 echinatis, connatis. An. *Ruellia?*
373. CELSIA *racemoſa.* - - Mm. - Barah. Flor. flavis, raçemoſis.
374. (*) BROVALLIA *humifuſa.* Ms. - Böka.
375. (*) LIMOSELLA *calycina.* Ma. - Uadi Zebîd.
376. (*) OROBANCHE *a) tinctoria.* - *Hôdar.* Môr. Srd. *Zybb elka* زب القع
377. - - *b) major ?* - - Taæs.
378. (*) CHARACHERA *a) tetragona.* Mi. *Frefrân.* Hadîe. *Mekatkata.* Blg.
379. (*) *b)* - *viburnoides.* - Mm. - Barah.
380. (*) SESAMUM *indicum:* - CP. *Djyldjylan.* Ubique.
381. (*) RUELLIA *a) imbricata.* Ms. - Boka.
382. (*) - *b) intruſa.* - - Mm. *Suudvud.* سودود Blg.
383. (*) - *c) ſtrepens.* - - Ms. *Ghobbâr.* غبار *Koſſeif.* قصيف *Meakeka.* منقنقه Srd.
384. - - *d) ſtrep. cærulea.* - Mi. - Srd. Flore cæruleo, parvo.
385. (*) - *e) guttata.* - - Ms. *Ghobîre.* غبيرة *Kasr.* كسر Djöbla. Chadra.
386. - - *f) alba.* - - - - Aludje. Flore albo, parvo; foliis acutis.
387. - - *g) hiſpida.* - - - *Chommâh.* خماح Ulif. Caule hiſpido perpendiculari.
388. (*) - *h) adhærens.* - - Mi. *Hamſched.* حمشد Hadîe.
389. (**) SIMBULETA. - Ms. *Symbulet ennefem.* سنبلة النسم Djöbla *Saſal.* سوسل
390. ACANTHUS *a) maderaſp ?* Mm. Mi. Ph. *Sabak.* سبك Srd. Floribus albis; bracteis, (non calyce) ciliatis.
391. (*) - *b) edulis.* - - Ps. Ph. *Zagaf.* Môr. Lhj.
392. (*) - *c) arboreus.* - - Mm. *Senaf.* سنف Blg. Barah.

(p 2) TETRA-

Fig. 5. Page CXV in one of the sections in the floristic part of the *Flora Aegyptiaco-Arabica*, listing species in *Flora Arabico-Yemen.* The names of the new species are marked with one asterisk; the new genus *Simbuleta* is marked with two asterisks but has no specific epithet. It was collected near Djöbla [Jibla]. *Ms* stands according to a list of abbreviations on page CI for *Montium regio superior* [higher region of the mountains]. The other abbreviations stand for: *Mi*: *Montium regio inferior* [lower region of the mountains]. *Mm*: *Montium regio media* [middle region of the mountains]. *Ma*: *Montium regio aquosa* [watery region of the mountains]. *CP*: *Cultæ in Planitiae* [cultivated in the plains]; *Ps*: *Planitiae argillaceae sicca* [dry, sandy plains]. Vernacular names and abbreviated names of localities are indicated to the right, sometimes followed by a few diagnostic words.

CXXI

MONOECIA.

531. LEMNA. - Ma. *Bleſemmân.* نلبسهان Uhf.
532. (✱✱) KOSARIA *fœtida.* - Mm. *Koſar.* قصر Hadîe.
533. ZEA *mays.* - - CM *Rumi.* Uhf.
534. (✱) PHYLLANTHUS *a) niruri.* Mi. Mm. *Mekatkata. Meneckete.* Blg.
535. (✱) - *b) maderaſp.* - Mi. Hadîe.
536. (✱) - *c) ovalifolius.* - Mm. *Homæmer.* حميمر Blg.
537. (✱) - *d) hamrur.* - Mm. *Hamrûr.* حمروري Melhân.
538. (✱) BUXUS *dioica.* – Mm. *Katam.* Kurina. Barah.
539 (✱) URTICA *a) palmata.* Mm. *Schadjaret el mehabbr. Hörokrok.* حرقرق Blg.
540. (✱) - *b) iners.* - - Mm. *Hamſched.* حمشد Blg.
541. (✱) - *c) paraſitica.* - Mm. *Nædjaa.* Blg.
542. (✱) - *d) divaricata.* Mi. *Mebærreka.* Hadîe.
543. - - *e) urens?* - Mm. *Kolahleba.* Blg. Capitulis albis axillaribus.
544. . - *f) dubia?* - Mm. Foliis oppoſitis, longitudine petioli, ovato-acutis, profunde ſerratis. Barah.
545. XANTHIUM *ſtrumar.* - Ma. *Mandj el mâ.* منج الما *Kavar el abîd.* Uhf. Môr.
546. AMARANTHUS *a) blitum.* Ph. *Schedacb.* Ghomfude. Môr.
547. - - *b) oleraceus.* - - CP. *Schedacb hindi.* Kudmîe.
548. - - *c) hybridus.* . *Schadjaret erraaf.* شجرة الرعاف Zebîd.
549. - - *d) racemis viridibus.* CP. *Aſîb el farras.* عسيب الفرس Zebid.
550. (✱✱) TURIA *a) ſativa.* - CP. *Turia.* Môr.
551. (✱) - *b) foliis cordatis.* - CP. Idem nomen.
552. (✱) - *c) leloja.* - . Ph. *Lua.* لوع Lhj.
553. (✱) - *d) gijef.* - Ph. Môr. Srd. *Gijef.*
554. (✱) - *e) moghadd.* - - Ph. Môr. *Moghadd*
555. (✱) JATROPHA *pungens.* Mi. *Mebarkeka.* Hadîe. Auſtraliorib. *Humejta.*
556. RICINUS *communis.* - Ph. Mm. *Tehſcha.* تتشع Hadîe. *Zajt.* Kudmie. *Djar.* Môr.
557. (✱) ACALYPHA *a) fruticoſa.* Ph. Mi. *Börtom.* برطم Srd. *Dafrâu.* Hadîe.
558. (✱) - *b) decidua.* - - Ph. Mi. *Bortom ſagbajar.* Ibid.
559. (✱) - *c) ſpicata.* - Ph. Srd.
560. (✱) - *d) ſupera.* - . Ph. Ibid.
561. (✱) - *e) ciliata.* - - Mi. Taæs.
562. (✱) CROTON *a) lobatum.* Ph. *Öhâb.* عبب Srd. *Medjerſche.* Môr.
563. - - *b) tinctorium?* Ps. *Hadak.* Lhj.
564. (✱) - *c) variegatum.* - Mm. *Dundul.* دندل Zebîd.

(q) 565.

Fig. 6. Page CXXI in one of the sections in the floristic parts of the *Flora Aegyptiaco-Arabica*, listing the species in *Flora Arabico-Yemen*. The new species in previously described genera are marked with one asterisk. *Kosaria* has two asterisks, indicating that it is a new genus with a new species. The first of the new species in the genus *Turia* is marked with two asterisks, indicating that it is a new genus with a new species. The remaining species in *Turia* have one asterisk each to mark that they are new species. For some species there are a few diagnostic words to the right of one or more vernacular names or localities.

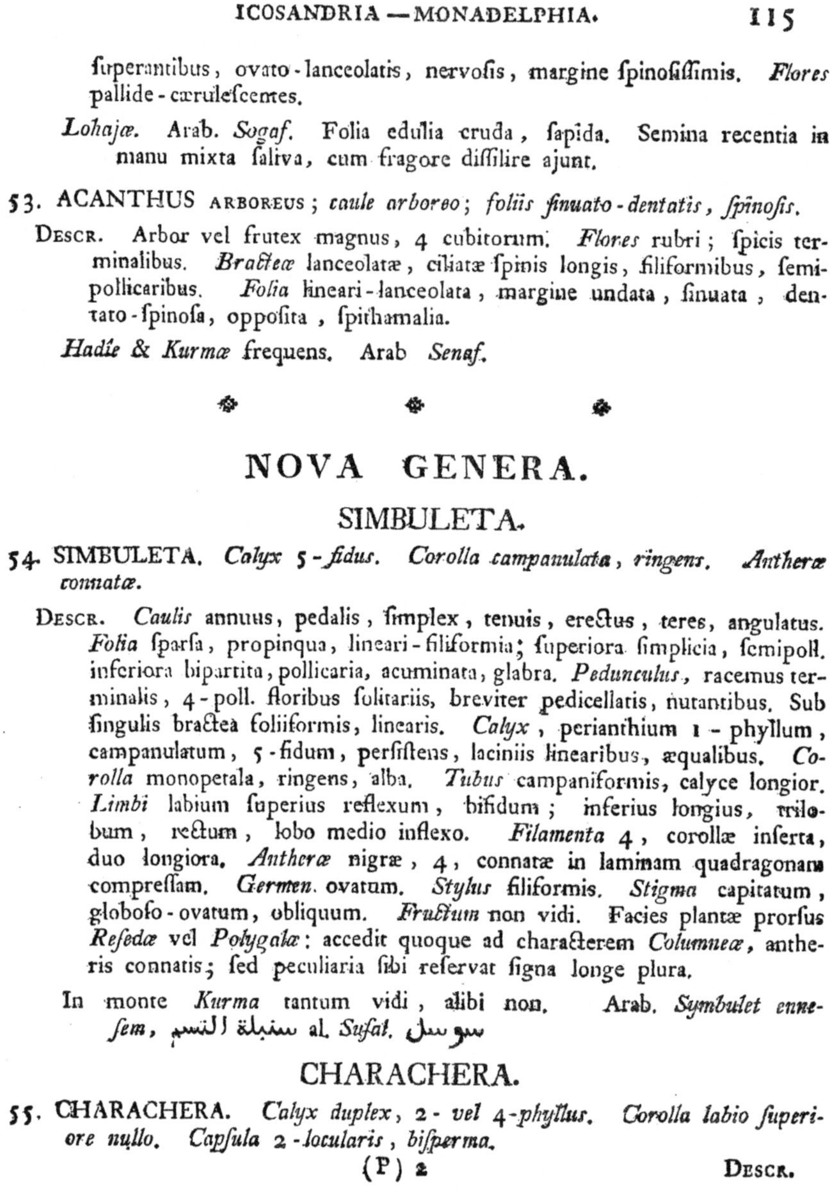

ICOSANDRIA — MONADELPHIA. 115

ſuperantibus, ovato-lanceolatis, nervoſis, margine ſpinoſiſſimis. *Flores* pallide-cæruleſcentes.

Lohajæ. Arab. *Sogaf.* Folia edulia cruda, ſapida. Semina recentia in manu mixta ſaliva, cum fragore diſſilire ajunt.

53. ACANTHUS ARBOREUS; *caule arboreo; foliis ſinuato-dentatis, ſpinoſis.*

DESCR. Arbor vel frutex magnus, 4 cubitorum. *Flores* rubri; ſpicis terminalibus. *Bracteæ* lanceolatæ, ciliatæ ſpinis longis, filiformibus, ſemipollicaribus. *Folia* lineari-lanceolata, margine undata, ſinuata, dentato-ſpinoſa, oppoſita, ſpithamalia.

Hadîe & *Kurmæ* frequens. Arab *Senaf.*

NOVA GENERA.

SIMBULETA.

54. SIMBULETA. *Calyx* 5-*fidus.* *Corolla campanulata, ringens.* *Antheræ connatæ.*

DESCR. *Caulis* annuus, pedalis, ſimplex, tenuis, erectus, teres, angulatus. *Folia* ſparſa, propinqua, lineari-filiformia; ſuperiora ſimplicia, ſemipoll. inferiora bipartita, pollicaria, acuminata, glabra. *Pedunculus*, racemus terminalis, 4-poll. floribus ſolitariis, breviter pedicellatis, nutantibus. Sub ſingulis bractea foliiformis, linearis. *Calyx*, perianthium 1-phyllum, campanulatum, 5-fidum, perſiſtens, laciniis linearibus, æqualibus. *Corolla* monopetala, ringens, alba. *Tubus* campaniformis, calyce longior. *Limbi* labium ſuperius reflexum, bifidum; inferius longius, trilobum, rectum, lobo medio inflexo. *Filamenta* 4, corollæ inſerta, duo longiora. *Antheræ* nigræ, 4, connatæ in laminam quadragonam compreſſam. *Germen* ovatum. *Stylus* filiformis. *Stigma* capitatum, globoſo-ovatum, obliquum. *Fructum* non vidi. Facies plantæ prorſus *Reſedæ* vel *Polygalæ*: accedit quoque ad characterem *Columneæ*, antheris connatis; ſed peculiaria ſibi reſervat ſigna longe plura.

In monte *Kurma* tantum vidi, alibi non. Arab. *Symbulet ennesem*, سنبلة النسم al. *Sufal.* سوفل

CHARACHERA.

55. CHARACHERA. *Calyx duplex*, 2-vel 4-*phyllus.* *Corolla labio ſuperiore nullo.* *Capſula* 2-*locularis*, *biſperma.*

(P) 2 DESCR.

Fig. 7. Page 115 in *Descriptiones*, *Centuria* IV, of *Flora Aegyptiaco-Arabica*, with protologue of the new genus *Simbuleta* and the beginning of the diagnosis of the genus *Characher*a. The protologue of *Simbuleta* consists of the new genus name, a diagnosis, a description, and notes on the locality: "In monte Kurma tantum vidi, alibi non." [Seen only in the mountains of Kurma, elsewhere not]. Arabic vernacular name *Symbulet ennesem* or *Sufal*. There is no name or description of a species.

164 CENTURIA VI.

RICINUS.

33. RICINUS MEDICUS; *foliis peltatis, 7-lobis; lobis lanceolatis, ſerratis, infimis magis divergentibus.*

DESCR. Folia interdum 8-loba. *Glandula* una ſimplex vel didyma, in apice petioli ſub folio, baſi ſtipularum loco utrinque, vel interdum ibidem una in medio etiam alæ. *Truncus* craſſitie ſemiſpith. Altitudo ultra 5 ulnas.

Roſettæ; colitur ob uſus medicos. Arab. *Charua.* خروع

❖ ❖ ❖

NOVA GENERA.

KOSARIA.

34. KOSARIA. *Calyx communis, monophyllus, carnoſus, 10-fidus. Receptaculum planum. Flores verruciformes, conferti.*

CHARACTER GENERICUS. *Calyx* communis, 1-phyllus, carnoſus, ſubtus convexus, carinis obtuſis. Limbus 10-12-fidus, radians, laciniis lineari lanceolatis, planis, craſſis, patenti-recurvis. *Receptaculum* placentiforme, ſemipoll. diametri, tegumento contiguo primum convexo, dein plano, germina tegente; inter quæ undique *pulpa* carnoſa. *Flores* numeroſiſſimi, velut verrucæ parvæ, in cruſta tegente recept. maſc. & fœm. mixti. *Maſculi*: *corolla* verruca conico-truncata, apice aperta, parva, viridis. *Filamenta* 2, raro 3, filiformia, albida, corolla longiora. *Piſtilla* nulla. *Fœminei* corolla verrucula in cruſta tegente recept. conica, clauſa, viridis: *Germen* ovatum, nitidum, album, in pulpa recept. reconditum: *Stylus* filiformis, albus: *Stigma* acutum. *Pericarpii* loco recept. in cujus pulpa deliteſcunt ſemina, quibus maturis dehiſcit apex corollæ fœm. *Semina* ſolitaria, ovato-trigona, acuta, apice carinata, nitida, albida.

SPECIES. *Caules*: ſpithamales, erecti, digitum craſſi, baſi ramoſi; *tuberculis* plurimis, ſparſis, confertis, truncatis, qui cicatrices ſunt foliorum delapſorum. *Folia* ſparſa; inferiora decidunt; ad apices reſtant. *Petioli* cylindrici, glabri, erecti, ſemipoll. & ſi in umbroſo creſcit 2 poll. *Lamina* lanceolata, ſæpe deflexa, ultra 2 poll. margine undulata, ſubdentata, glabra, membranacea, viridis; interdum rufeſcens, baſi cordata. *Pedunculi* axillares, ſolitarii, recurvi, teretes, petiolo breviores. *Calyx* communis ſæpe pendulus, interdum verticalis, raro pollicaris diametri.

GYNANDRIA — POLYGAMIA. 165

diametri. Laciniæ limbi calycini æquales, ſæpe alternæ, minores, in eadem planta. *Floſculi* ut parvæ verrucæ e recept. exſtantes: *maſculi* filamentis antheratis: *fœminei* ſtylo capillari acuto; ſubtiliſſimi.

Hadîe in Coffeæ-cetis ſub arboribus inter lapides. Arab. *Koſar.* قصر
Planta lacteſcens, fœtens; flore *Marchantiam* refert. Contuſa *exanthemata* curat. Sale mixta datur Vaccis ægris, gravidis.

Fig. 8. Page 164 and first paragraph of Page 165 in *Descriptiones*, *Centuria* VI, of *Flora Aegyptiaco-Arabica*, with protologue of the new genus *Kosaria*. The protologue consists of the new genus name, a diagnosis, a description of the genus, a description of the single species and notes on habitat and locality: "Hadie in Coffeæ-cetis sub arboribus inter lapides." [Under trees and between rocks in the coffee-plantations near Hadie.]. The Arabic vernacular name *Kosar*. There is a separate description of a species, but no species name. For the full species name, *Kosaria foetida*, see p. CXXI, No. 532, in the floristic part of the book (Fig. 6).

Vahl (1790, 1791, 1794, 1804, 1805) carefully studied Forsskål's publications and herbarium and, as seen in this work, he often provided solutions to the issues mentioned above. Unfortunately, Vahl did not follow the rule of priority, which was only set out as an important principle in botanical nomenclature by Candolle (1813, 1819) in his *Théorie élémentaire de la botanique*. Vahl has therefore sometimes renamed Forsskål's names even when they should have priority according to the present ruling of the ICN.

As explained in the chapter above, *The letter 'å' in Scandinavian …*, the letter was used at least since the 15th Century in Sweden but at first hardly known or available in typography outside Sweden. For publications with loose types printed outside Scandinavia, the letter used to represent 'å' depended on what typeface was available to the typesetters, or it was simply printed as 'a'. Moreover, as shown in the chapter *The spelling of Forsskål's surname*, his name and the surnames of other members of his family have alternatively been spelt with a single or a double 's' and sometimes with an 'h' before the final 'l', but Forsskål himself always wrote it as 'Forsskål', at least when writing to Swedes (Fig. 1). Friis & Thulin (1984) recommended that the spelling 'Forsskål' should be used as a standard, and this is now generally accepted, with the abbreviation 'Forssk.' when Forsskål is the author of plant names. TL-2 (2024) has the name as 'Forsskål, Pehr (Peter)', and the standard abbreviation as *Forssk.*

Outside Scandinavia, attempts were made by typesetters *ad hoc* to modify the letter 'a', or rarely the letter 'o', with a small 'o' or a small ring above it, an 'overring', or sometimes the lower part of an 'o' or the lower part of a 'u' above it. Examples of attempts to make 'å' with rings or half-circles can be seen in Fig. 9 (an 'o' or a ring over an 'a'), Fig. 10 (the lower part of an 'o' or of a 'u' over an 'a'), and Fig. 11 (a ring over an 'o'). Sometime eponyms based on Forsskål's name have been printed with 'a' with a circumflex where the name has an 'å'. Since the 'â' in Forsskål eponyms is clearly not the French letter 'â', we have here interpreted these attempts to represent an 'å' as indications of the author's intention, and have interpreted such eponyms as containing an 'å'.

Later authors have named or renamed 73 of Forsskål's names of plants with eponyms, which is more than one tenth of all Forsskål's new species of plants. In the original literature publishing new names, Forsskål's eponyms can be found in 34 variants, listed below in a declining sequence of occurrences (distinguishing between variants with the vowels 'â' and 'å', but not between eponyms with uppercase and lowercase initial letter). The variants *forskalii* and *forskahlii* are the most common, both occurring in the original publications more than 10 times. The numbers in brackets indicate the number of that particular variant: *forskalii* (17); *forskahlii* (12); *forskahlei* (5); *forskåhlii* (5); *forskålii* (5); *forskålei* (3); *forskohlei* (3); *forskohlii* (3); *forskålæi* (2); *forskåli* (2); *forskaliana* (2); *forskaolii* (2); *forskoehlii* (2); *forskolei* (2); *forsk.* [sic!] (1); *forskaehlei* (1); *forskailii* (1); *forskal.* [sic!] (1); *forskâlaei* (1); *forskålaei* (1); *forskalei* (1); *forskâlei* (1); *forskålianum* (1); *forskalianus* (1); *forskaoli* (1); *forskaolii* (1); *forskoeli* (1); *forskoelii* (1); *forskohleana* (1); *forskoklii* (1); *forskölei* (1); *forskolii* (1); *forsskaoliana* (1); *forsskaolii* (1); *forskaolii* (1).

The main part of this work deals with Forsskål eponyms under the botanical code, but both plants and animals have been named with scientific names commemorating Forsskål. On the expedition, he collected and studied specimens of both plants and animals and drafted manuscripts on both, and like the botanists, zoologists have also coined eponyms based on Forsskål's surname. However, the rules in the *International Code of Zoological Nomenclature* (ICZN; International Commission on Zoological Nomenclature 1999) differ from the rules in the *International Code of Nomenclature for Algae, Fungi and Plants* (ICN) in the ruling in Art. 32, where, according to Art. 32.3, the correct original spelling of names for animals has to be preserved unaltered, except where it is mandatory to change the suffix or the gender ending, and where there are incorrect spellings. The latter is defined in ICZN, Art. 32.5.1 and 32.5.11, as cases where there is clear evidence of an inadvertent error, such as a *lap sus calami* or a copyist's or printer's error [equivalent to the typographical error in the *International Code for Algae, Fungi and Plants*]. These cases must be corrected. Incorrect transliterations or Latinizations are not to be considered inadvertent errors. Correction of a spelling of a name in a publisher's or author's *corrigendum* issued simultaneously with the original work is to be accepted as clear evidence of an inadvertent error in the original text. Also unlike the ICN, the ICZN requires (in Art. 32.5.2.1) that in the case of a diacritic or other mark in the original spelling of a name, the mark concerned is deleted, except that in a name published before 1985 and based upon a German word, the *Umlaut* sign is deleted from a vowel and the letter 'e' is to be inserted after that vowel.

At the level of genus, there is under the ICN the name *Forsskaolea* L., a Forsskål eponym of a genus of plants originally published as *Forsskålea*. Under the Zoological Code there is a currently accepted genus of siphonophores named *Forskalia* Kölliker (1853: 2–10, Tab. I-II), spelt with a single 's' and 'å' transcribed to 'a'. *Forskalia* has probably five species distributed in the Atlantic, the Mediterranean, the Red Sea, and the Pacific (Pugh 2003; GBIF continuously updated), and is the only genus in the monotypic family Forskaliidae Haeckel (1888), in the order Siphonophorae Eschscholtz, 1829, of the phylum Cnidaria Hatschek, 1888. In Köllicher's original publication, there is no mentioning of why the genus was named *Forskalia*. According to Wolff (1990: 237) there is no clear link between this genus name and preserved material in Forsskål's zoological collections or species mentioned in his publications, and the name *Forskalia* for a pelagic siphonophore is therefore almost certainly honorific, based on Forsskål's reputation as a pioneer marine biologist studying the open seas. Haeckel proposed another genus with a new species, *Forskaliopsis magnifica* Haeckel, 1888, from the Indian Ocean, but there is no description of it, and it is treated as *nomen nudum* under *Forskalia*. Yaron *et al.* (1986) published a revision of the mollusc-names based on Forsskål's publications and collections, but they only mention one eponym, the genus name *Forskalena* (Yaron *et al.* 1986: 191, erroneously spelt *Forskaelena*). *Forskalena* Iredale (1918), listed in GBIF (continuously updated) and WoRMS (continuously updated), is the name of a new genus of snails identical with, but the genus name not a replacement for *Forskalia* H. & A. Adams (1854). H. & A. Adams' name was preoccupied by the name *Forskalia* Kölliker (mentioned above). The snail is now referred to the genus *Gibbula* Risso, 1826.

Fricke (2008) published an analysis of Forsskål's names for five new genera of fish (all still in common use) and 153 names for new species (90 of these still in common use), manly from the Mediterranean and the Red Sea. Fricke listed only ten Forsskål eponyms given to species of fish (the authorities of the names of animals are here given in the form proscribed in the Zoological Code and not listed in *Literature cited* in this work): *Atherina forskalii* Rüppell, 1837 [based partially on *Atherina hepsetus* Forssk., 1775]; *Atherinomorus forskalii* Rüppell, 1838 [also based partially on *Atherina hepsetus* Forssk., 1775]; *Cheilio forskalii* Valenciennes in Cuvier & Valenciennes, 1839 [a junior synonym of *Labrus inermis* Forssk.]; *Crenidens forskalii* Valenciennes in Cuvier & Valenciennes, 1830 [a junior synonym of *Sparus crenidens* Forssk., currently *Crenidens crenidens* (Forsskål, 1775)]; *Holocentrus forskael* Lacepède, 1802 [a junior synonym of *Perca (?) fasciata* Forssk., currently *Epinephelus fasciatus* (Forsskål, 1775)]; *Hydrocynus forskahlii* Cuvier, 1819 [based on Forsskål's misidentification of a new species as *Cyprinus dentex* Linnaeus, 1758, which now is named by the current name *Alestes dentex* (Linnaeus 1758)]. The name *Salmo roschal* Walbaum, 1792, was proposed on Forsskål's description, taking up a vernacular name cited by Forsskål as the epithet and thus antedating Cuvier's name; Fricke (2008) suggested that in order to preserve the use of *Hydrocynus forskahlii* Cuvier, a petition to the International Commission on Zoological Nomenclature was needed to suppress *Salmo roschal*; *Parupeneus forsskali* (Fourmanoir & Guézé, 1976) [described by Forsskål as *Mullus auriflamma* Forssk. that threatened its commonly used junior synonym *Upeneus vanicolensis* Valenciennes in Cuvier & Valenciennes, 1831, *Mullus auriflamma* Forsskål, 1775, was therefore placed on the Official Index of Rejected Names in Opinion 846 (Anonymous 1968). *Pseudupeneus forsskali* Fourmanoir & Guézé,

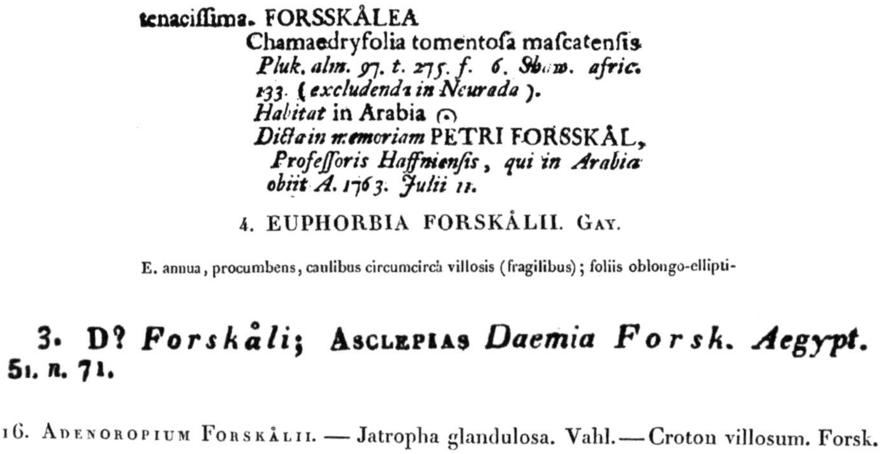
tenacissima. FORSSKÅLEA
Chamaedryfolia tomentosa mascatensis.
Pluk. alm. 97. t. 275. f. 6. Shaw. afric.
133. (excludenda in Neurada).
Habitat in Arabia ☉
Dicta in memoriam PETRI FORSSKÅL,
Professoris Haffniensis, qui in Arabia obiit A. 1763. Julii 11.

4. EUPHORBIA FORSKÅLII. Gay.

E. annua, procumbens, caulibus circumcirca villosis (fragilibus); foliis oblongo-ellipti-

3. D? *Forskåli*; Asclepias *Daemia Forsk. Aegypt.* 51. n. 71.

16. Adenoropium Forskålii. — Jatropha glandulosa. Vahl. — Croton villosum. Forsk.

Fig. 9. Examples of Forsskål eponyms with the letter 'å' represented correctly by an 'o' or a ring over an 'a'. No. 1 from above: *Forsskålea tenacissima* L., *Opobalsamum*: 19 [no text on a 'p. 18'; text on p. 19 follows immediately after the text on p. 17], 1764. (Urticaceae). This text was printed in Sweden. No. 2 from above: *Euphorbia Forskålii* J.Gay, in Webb & Berth., *Hist. Nat. Iles Canaries* (Phytogr.) 3 (pt. 2, sect. 3): 240, 1847 [Apr.–May 1837]. (Euphorbiaceae). This text was printed in Paris. No. 3 from above: *Daemia [Schultes's misspelling for Doemia]? forskåli* Schult., *Syst. Veg.* ed. 15bis, 6: 113, 1820 [Aug.–Dec. 1820]. (Apocynaceae). This text was printed in Stuttgart. No. 4 from above: *Adenoropium Forskålii* Pohl, *Pl. Bras. Icon. Descr.* 1: 15, 1826. (Euphorbiaceae). This text was printed in Vienna.

1976, was described as a replacement name for *Mullus auriflamma*, and the species has the current name *Parupeneus forsskali* (Fourmanoir & Guézé, 1976)]; *Pseudoscarus forskalii* Klunzinger, 1871; *Pseudupeneus forsskali* Fourmanoir & Guézé, 1976 [see above under *Parupeneus forsskali* (Fourmanoir & Guézé, 1976)]; *Trigon forskalii* Rüppell, 1829. This list indicates that the orthographic variant *forskalii* is the most common, occurring in six names, the other four names are each represented only with one variant. As for Forsskål's plant eponyms, the reasons for giving eponyms are most often complications with the interpretation of Forsskål's zoological work (Forsskål 1775b), with the result that a junior synonym has been published. For further information about these names, see Fricke (2008).

It is not possible in this work, focussing on plants, to analyse all Forsskål eponyms under the Zoological Code. However, a search in two zoological databases gave results that are comparable to those for plants with regard to the orthographic variety and the most frequent variants: In Sherborn (1902, 1922–1932), which has not recorded names later than 1932, 56 eponym epithets are listed with 19 orthographic variants, and WoRMS (continuously updated), which records marine animals, 115 eponym epithets are listed with 20 orthographic variants, here both recorded in a declining sequence of occurrences and with the number of records in brackets after each variant. [Sherborn]: *forskalii* (18), *forskahlii* (7), *forskaelii* (4), *forskalea* (4), *forskaleana* (4), *forskaliana* (3), *forskael* (2), *forskaolii* (2), *forskaeli* (1), *forskahli* (1), *forskal* (1), *forskalia* (1),

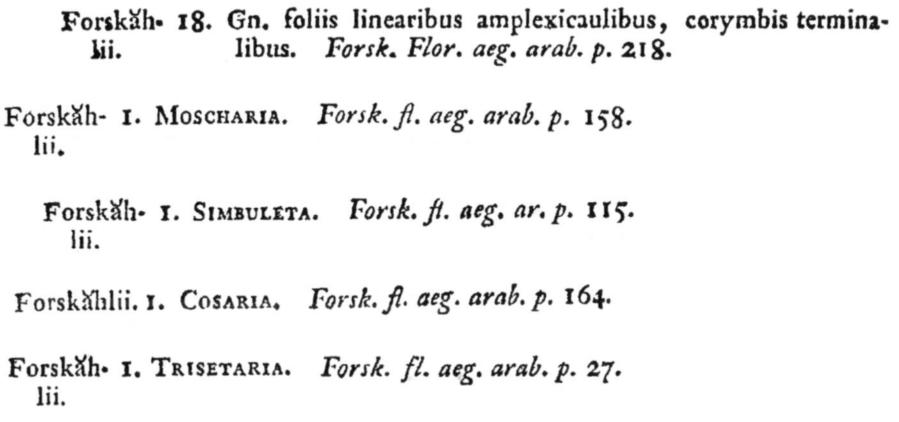
Forskåh- 18. Gn. foliis linearibus amplexicaulibus, corymbis terminalii. libus. *Forsk. Flor. aeg. arab. p.* 218.

Forskåh- 1. Moscharia. *Forsk. fl. aeg. arab. p.* 158.
lii.

Forskåh- 1. Simbuleta. *Forsk. fl. aeg. ar. p.* 115.
lii.

Forskåhlii. 1. Cosaria. *Forsk. fl. aeg. arab. p.* 164.

Forskåh- 1. Trisetaria. *Forsk. fl. aeg. arab. p.* 27.
lii.

Fig. 10. Examples of Forsskål eponyms with the letter 'å' represented by the lower part of an 'o' or 'u' over an 'a'. All from J.F.Gmel., *Syst. Nat.* ed. 13bis, 2 (1791–1792). No. 1 from above: *Gnaphalium Forskåhlii* J.F.Gmel., *Syst. Nat.* ed. 13bis, 2(2): 1214 (1792 [1791 publ. late Apr.–Oct. 1792]). (Asteraceae). No. 2 from above: *Moscharia Forskåhlii* J.F.Gmel., *Syst. Nat.* ed. 13bis, 2(1): 275 (1791 [late Sep.–Nov. 1791]). (Lamiaceae). No. 3 from above: *Simbuleta Forskåhlii* J.F.Gmel., *Syst. Nat.* ed. 13bis, 2(1): 242 (1791 [late Sep.–Nov. 1791]). (Plantaginaceae). No. 4 from above: *Cosaria Forskåhlii* J.F.Gmel., *Syst. Nat.* ed. 13bis, 2(1): 22, 71 (1791 [late Sep.–Nov. 1791]). (Moraceae). No. 5 from above: *Trisetaria Forskåhlii* J.F.Gmel., *Syst. Nat.* ed. 13bis, 2(1): 177 (1791 [late Sep.–Nov. 1791]). (Poaceae). These texts were all published in Leipzig.

forskalina (1), *forskallii* (1), *forskillii* (1), *forskoehlii* (1), *forskohlii* (1), *forskolea* (1), *forskåleana* (1).

In the eponyms listed in WoRMS the distribution of the variants are thus: *forskalii* (43), *forskali* (20), *forskaelana* (9), *forskoehlii* (8), *forskahlii* (7), *forskaliana* (6), *forskahli* (3), *forskalea* (3), *forskalana* (2), *forskalina* (2), *forskålii* (2), *forskael* (1), *forskaelii* (1), *forskalensis* (1), *forskalia* (1), *forskalli* (1), *forskallii* (1), *forskhalii* (1), *forskäli* (1), *forskalli* (1).

The frequency of the orthographic variants is rather different in the two zoological databases, but *forskalii* with one 's' and 'å' transcribed to 'a' is the most common in both.

Correcting or not correcting orthography of eponyms with 'å' and single or double consonants

We have studied three cases where nomenclatural committees under different International Botanical Congresses have decided upon orthographic corrections of names of new plant genera based on personal names with 'å' and single vs. double consonants. Two of the cases are Swedish, one deals with the Scandinavian 'å', another with single or double 'ss', and a third case is Italian, with a single or double 'zz'. The outcome of the first two cases were decided on between 1999 and 2016 by the Committee for Bryophytes, and the Committee for Fossils, and at the next higher level the final decisions were made by the General Commit-

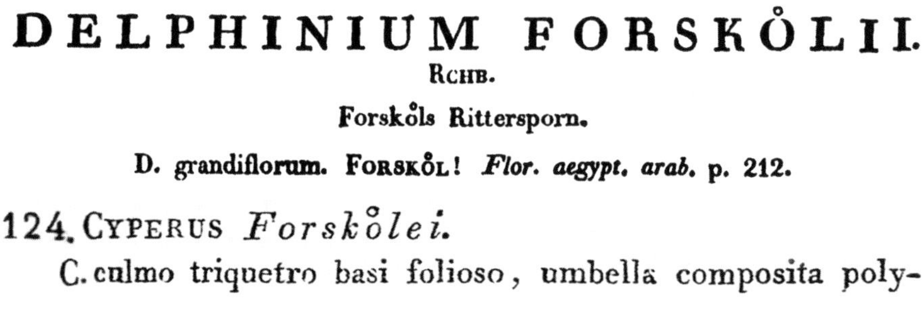

DELPHINIUM FORSKÔLII.
RCHB.
Forskôls Rittersporn.
D. grandiflorum. FORSKÔL! *Flor. aegypt. arab.* p. 212.

124. CYPERUS *Forskôlei.*
C. culmo triquetro basi folioso, umbella composita poly-

Fig. 11. Examples of Forsskål eponyms with the letter 'å' represented by an 'o' or a ring over an 'o'. Above: *Delphinium Forskålii* Rchb., *Ill. Sp. Acon. Gen.* 5: Tab. 68, 1823 (Ranunculaceae). This text was published in Leipzig. Below: *Cyperus Forskålei* A.Dietr., *Sp. Pl.* ed. 6, 2: 251, 1833 (Cyperaceae). This text was published in Berlin.

tee. Here, we have summarised these decisions and discuss what consequences there may be for plant names based on Forsskål's personal name.

Aongstroemia Bruch & W.P.Schimper (1846) [original spelling Angstroemia], nom. cons. (Mosses; Bryophyta).

Aongstroemia Bruch & W.P.Schimper in Bruch, Schimper & Gümbel (1846: separately paginated chapter named *Angstroemia*, p. 1; Facsimile edition vol. 1: 171). According to Bruch and Schimper, the name of this moss genus commemorates "unserem Freunde, dem eifrigen Bryologen A. Angström [in fact, Johan Ångström (1813–1879)] zu Upsala" [our friend, the eager bryologist A. Angström in Uppsala]. In spite of the initial of the first name, this must refer to the medical doctor Johan Ångström, a recognised bryologist, and not to his brother, the more widely known physicist and astronomer Anders Jonas Ångström (1814–1874), for whom the unit of measurement 'Angstrom' or 'Ångström' has been named. A proposal, no. 1175 (Margadant & Geissler 1995: 613) to the nomenclature committees suggested rejection of *Angstroemia* against *Aongstroemia* because of the common use in newer literature of the latter orthography. The Committee for Bryophyta (Zijlstra 1998) concluded that the spelling *Angstroemia* was not a typographical error as defined in the ICN, Art. 60.1, because the Swedish letter 'Å' was systematically transcribed as 'A' in the works of Bruch & Schimper. The Committee for Bryophyta considered a transliteration of 'å' to 'a' a deliberate Latinization and did not discuss if the use of 'a' instead of 'å' could be an orthographic error, correctable according to ICN, Art. 60.1. The name *Angstroemia* therefore, according to the Committee for Bryophyta, did not fall under ICN, Art. 60.7, and it was not required that 'a' should be corrected to 'å', which again should be transliterated to 'ao'.

However, an orthographic variant of the name of the moss genus with an initial letter 'Å', as *Ångströmia*, had been used by Paris (1894: 34), and many or most recent authors had in fact gone back to the personal name from which the plant name had been derived, 'Ångström', and then transcribed the corrected name as *Aongstroemia* in agreement with ICN, Art. 60.7. This orthographic variant was accepted in general handbooks such as Wijk (1959: 81) and Greuter *et al.* (1993: 67). The Committee for Bryophyta unanimously recommended conservation of the orthography as *Aongstroemia* due to the wide use of this orthography in literature, and the recommendation was accepted by the General Committee (Nicolson 1999) and the subsequent XVI International Botanical Congress in St. Louis, USA, in 1999.

There are other moss names dependent on the accepted spelling of *Aongstroemia*: A family name, Aongstroemiaceae De Not. (Notaris 1869: 30) [original spelling Angstroemiacei], which in agreement with the conservation of the generic name is corrected to Aongstroemiaceae De Not. A genus distinct from *Aongstroemia*, was named *Aongstroemiopsis* M.Fleischer (1904, as 1900–1902: 331) [original spelling as '*Angstroemiopsis*' in the key, and *Ångstroemiopsis* in the protologue]; the generally accepted orthography is now 'Aongstroemiopsis', but no formal decision by nomenclatural committees has been taken on this. *Protoaongstroemia* Fedosov *et al.* (2023: 21) [original spelling the same] is a new genus published with a new species (*Protoaongstroemia sachalinensis* Fedosov *et al.*, gen. et spec. nov; Fedosov *et al.* 2023: 21), the name is validated with a *descriptio generico-specifica*, in agreement with ICN, Art. 38.5.

We accept that the plant name derived from the personal name Ångström had in the original publication been Latinized with an initial 'A', but because of the common use of the variant *Aongstroemia*, it can and has been conserved with that spelling. It is not difficult to draw the consequences of this case for our treatment of Forsskål's eponyms. The use of 'a' instead of 'å' in original publications of plant names means that 'a' is also to be used in current names without first correcting to 'å'. The use of 'å' or attempts to print 'å' in original publications of plant names means that the 'å' has to be transliterated to 'ao'.

Nilssonia A.T.Brongn. (1825, as '1824': 210), [original spelling Nilsonia].

Brongniart (1825, as '1824') described the formation of the Höör sandstone in central Scania, and continued "C'est là M. Nilson, professeur à Lund, a observé pour la premièr fois … restes de végétaux …" [It is there that Mr. Nilson, professor at Lund, [the Swedish zoologist and palaeontologist Sven Nilsson (1787–1883)] for the first time has observed the plant remains …]. Later, Brongniart had also collected material of a plant fossil, a Cycad, and described it. The fossils represented "un genre particulière dans la famille des Cycadées, et nous les donneront le nom de *Nilsonia*, en l'honneur du savant professeur …" [a distinct genus in the family of Cycadaceae, and we give them the name *Nilsonia* in the honour of the learned professor].

Wang (2011: 240), in proposal 1995 to the nomenclature committees, noted that the two variants *Nilsonia* and *Nilssonia* were alternately used by palaeobotanists for the same genus of fossil foliage of Mesozoic cycadophytes, and that recently it had been proposed to revert to *Nilsonia*, Brongniart's original spelling. Pointing out that the plant was named after Sven Nilsson, but because Brongniart continued to use the spelling *Nilsonia* in his original and subsequent works, this was not a typographical error. Since the 1830s, the spelling *Nilssonia*, ascribed to Brongniart, but with the spelling as in Nilsson's surname, had been widely used in palaeobotanical literature. Wang therefore proposed conservation of the widely used spelling *Nilssonia* under Art. 14.11, which allows conservation to preserve a particular spelling or gender.

Herendeen (2015: 1306) reported from the Committee on Fossils that although two members of the Committee were not convinced that reverting to the original spelling would be disruptive, a strong majority (13–2–0) voted in favour of recommending its acceptance. However, the General Committee (Wilson 2016: 2051) considered Brongniart's original spelling to be a correctable error: "The GC rejected (4–21–0–0) prop. (1995) to conserve *Nilssonia* with that spelling as being unnecessary." This was subsequently approved by the XIX International Botanical Congress in Shenzhen, China, in July 2017.

This case is the basis of ICN, Art. 60.1. Ex. 7, an example of a correctable orthographic error. Having demonstrated in the chapter *The spelling of Forsskål's surname* that Forsskål consistently spelt his name with double 's', it is necessary to decide if, when Forsskål eponyms spelt in the original publication of eponyms with one 's', this should be corrected to an orthographic variant with 'ss'. One of our reviewers called special attention to ICN, Art. 60.1. Ex. 7, stating that this Example is "directly relevant to the topic. This Example demonstrate that if the name of a person was misspelt in the protologue (for whatever reason), the resulting plant name is correctable in accord with the established spelling of the one person's name." Our other reviewer disagreed: "One must also note that there exists official variants [of Forsskål eponyms] with one or two s … which should not lead to correction."

This leaves us with a difficult choice. Since the vast majority of the Forsskål eponyms are spelt with one 's' in the original publications, including a relatively recent name as *Cyperus forskalianus* Väre & Kukkonen (2005), and only two eponyms are spelt with double 'ss' in the original publication (*Maytenus forsskaoliana* Sebsebe, 1985, and *Ruellia forsskaolii* Thulin, 2016), we are aware that changing to double 's' in Forsskål eponyms would cause at least a short-term nomenclatural disruption, probably also a longer-term confusion, and that it can always be claimed that the spelling with one 's' is an intentional Latinization.

The decisions taken in the *Nilssonia*-case by the General Committee and at the XIX Botanical Congress was based on the knowledge that about half of the cases where the name was used accepted the spelling with double 's', the other half used single 's'. One of our reviewers has also pointed out that: "Another important matter is that [ICN,] Art. 60 offers some reservations in such corrections but (in the examples) they are limited to certain established generic names …, and the Article ultimately appeals to conservation if the preservation of the incorrect or irregular spelling is desirable [Note 1, Ex. 9 and 10)." We find that rigid and logical correction of orthographic variants of Forsskål epithet eponyms with one 's' to variants with 'ss', although in agreement with the way Forsskål spelt his name, would be against ICN, Art. 60.3 "The liberty of correcting a name is to be used with reserve …", particularly because it will go against the dominant use of orthographic variants with one 's'. See also the discussion below of *Albizia versus Albizzia*.

Albizia Durazz. (Durazzini 1772).

In their discussion of when and when not to correct a plant name to agree with the name of a person who is honoured, Brummitt & Taylor (1990: 304–305) cited the case of *Albizia* Durazz. (1772), the name of a widespread genus of leguminous trees, as an example of contrasting views through history with regard to correction. The genus was named after Filippo degli Albizzi, who in eighteenth century Florence had introduced a species of that genus as an ornamental tree (Targioni-Tozzetti 1853: 241). Bentham (1844: 84) took up the name but corrected it to *Albizzia*, in accordance with the correct spelling of the name on which the eponym was based. Bentham's corrected spelling persisted until the mid-1900, when Little (1945: 510) corrected the name back to *Albizia*, having noted that Durazzini consistently used the single 'z' spelling in the original publication and pointed out that the orthographic variant *Albizia* could therefore not be a typographical error (he did not discuss if it could be an orthographic error). This view gained acceptance in floristic works, probably assuming that Durazzini's spelling with one 'z' was an intentional Latinization of Albizzi's name. If so, it should not be corrected according to ICN, Art. 60.9 (protecting intentional Latinizations, except when dealing with terminal vowels in personal names).

However, *Albizia* might also be held to be an orthographic error correctable to *Albizzia*, as was *Nilsonia* to *Nilssonia* in the example above. A correction to *Albizzia* would perhaps now seem to be against the common use of the spelling *Albizia*, as recorded by Greuter *et al.* (1993: 30), IPNI (continuously updated) and POWO (continuously updated). *Albizia* Durazz. is also the accepted spelling in the ICN, Art. F.91, Ex. 1, where the name of a fungus, *Meliola albizziae* Hansf. & Deighton is formally corrected to *M. albiziae* "in accordance with the accepted spelling of *Albizia* Durazz."

We have not been able to find a clear-cut or unmistakeable ruling in the ICN that allows us to determine when an orthographic variant is meant to be an intentional Latinization and when it may represent an orthographic error, and neither *Albizia*, nor *Albizzia* is included in Appendix III of the ICN, the list of *Conserved, Protected, and Rejected Names of Genera and Subdivisions of Genera*. It seems that *Albizia* is the most common orthographic variant in taxonomic literature, while *Albizzia* is quite common in horticultural literature.

We find that retaining the spelling *Albizia*, as in the original publication, for the name of the legume tree derived from the personal name Albizzi, does contradict the General Committee's decision to correct *Nilsonia* to *Nilssonia*. However, because *Albizia* is in common use, we find that ICN, Preample 13 "In the absence of a relevant rule or where the consequences of rules are doubtful, established custom is followed", and *Albizia* should therefore not be corrected to a variant with double 'zz'. Eventually, conservation could settle the question. The question of whether *Albizia* should be spelt with 'z' or 'zz' is not absolutely analogous to the question of whether Forsskål's eponym epithets should be spelt with 's' or 'ss', and we will draw our conclusion in the text below, remembering that few epithets honouring Forsskål with 'ss' are in use or were the spelling in the original publication.

Our two anonymous reviewers of this work have offered partly opposite views on how to interpret the ICN with regard to correction of orthographic variants. Our first reviewer concludes:

> My opinion is thus ... that most original spellings of names and epithets dedicated to Forskal [sic!], with the Code [ICN] as it stands, cannot be changed [we take this as meaning 'cannot be corrected'], with the ... exception of those where a diacritical sign [å] is transcribed [to 'a' with] an added 'o' ... and those rules about terminations in Art. 60.8 that imposes -ii.

We find that the case cited above of *Angstroemia* having to be conserved in order to be correctly spelt *Aongstroemia* because the original publication consistently spelt it with an 'A' agrees with the view of our first reviewer. We also cited above the first reviewer's opinion: "One must also note that there exists official variants [of Forsskål eponyms] with one or two s ... which should not lead to correction", and we discussed in the previous chapter the two cases of *Nilsonia* versus *Nilssonia* and *Albizia* versus *Albizzia*, concluding that, in agreement with ICN, Preamble 13, when the rules are doubtful, established customs should be followed, which is to use only one 's' in the Forsskål eponyms where this was the spelling in the original publication. We have therefore not corrected the many orthographic variants of Forsskål eponyms with one 's' to variants with 'ss'.

As mentioned in our chapter *The spelling of Forsskål's surname*, a number of members of the Forsskål family inserted an 'h' in their surname after the 'å', including Pehr Forsskål's brother Johan Christian Forsskåhl in correspondence with Linnaeus, but this was not used by Pehr Forsskål himself.

Linnaeus himself used the spelling Forskåhl when writing the surname of Pehr Forsskål in *Mantissa Plantarum* in 1767, and this spelling was followed by Gmelin in his edition of *Systema Naturae* in 1791 and by many other later authors. This is the decisive reason why we accept the use of an 'h' in the original spelling of the eponyms. We may also mention that our first reviewer remarked on this that "adding an h [in the eponym in the original publication] is intentional Latinization protected by Art. 60.9 and that it thus cannot be changed ..." We agree that the addition of an 'h' has to be accepted in Forsskål eponyms, but rather because of the existence of these alternative spellings of the surname than because we accept it as an intentional Latinization.

The second reviewer acknowledges the rules of the ICN but looks forward to a more radical, but logical standardisation, probably by future changing of the ICN:

> Typically, eponyms honouring the same person are spelled invariably [in a standardised way] ... I suggest ... a greater emphasis on the spelling of Forsskål's name and its consequences to the application of [ICN,] Art. 60, and to adopt a single legitimate spelling variant (*forsskaolii*, etc.) that could be the basis for standardisation. ... No doubt, this [the correct orthography of

> plant names] is still the most debatable matter [about the rulings in the ICN], which is subject to opinions and is open to interpretations. As the current nomenclatural rules are ambiguous and allowing for interpretations, this practically means that there may be no final 'truth' in such studies when seen from either side, the authors or reviewers, but both sides develop experience in trying to evaluate the subject. I consider that such studies necessarily contribute not only directly to the subject, but also to developing the rules of botanical nomenclature.

We cannot reconcile the views of our two reviews about whether to correct or not to correct. However, ICN, Art. 61.1 allows only one orthographical variant of any one epithet as validly published. Therefore, we cite in the systematic part of this work each eponym as originally published and as corrected in agreement with the rules of the ICN. This orthographic variant, in agreement with ICN, Art. 60, may differ from what would be the epithet as formed in agreement with ICN, Recommendation 60C.3, from the correct spelling of Forsskål's surname. The recommendation states that: "In forming new epithets based on personal names the customary spelling of the personal name should not be modified unless it contains letters, ligatures, or diacritical signs ..." Using the spelling Forsskål as a basis, this would lead to only four variants of the eponym, *forsskaolii*, *forsskaolianus*, *forsskaoliana* and *forsskaolianum*, and this limited range of variants would bring Forsskål's eponyms in line with most other eponyms. However, for this to happen a change of the ICN would be required, and a sequence of proposals made to the XX International Botanical Congress in Madrid, 2024 (Demoulin 2023) towards standardisation of epithets were rejected at the Nomenclature session, and we are therefore left with the existing rules of Art. 60–61 (Turland *et al.* 2024, except for minor amendments mentioned below).

In order to decide if the vowel with a diacritic sign in Forsskål's eponyms should be corrected, we have had to decide when the typographers' *ad hoc* typography to print an 'å' really does represent an 'å'. There is no rule in the ICN about this, but since these *ad hoc* letters, as well as 'å', have been used in Forsskål eponyms in the attempt to represent 'å', we have decided to treat them as that letter 'å'.

For this work, we have followed the latest published version of the ICN, the *Shenzhen Code* (Turland *et al.* 2018). However, two proposals to modify the *Shenzhen Code* (Turland *et al.* 2018), relevant to Forsskål's eponyms, were made for the XX International Botanical Congress at Madrid, 2024, and were accepted by the Congress (Turland *et al.* 2024).

McNeill *et al.* (2022: 1341, Proposal 187, by the Rapporteurs referred to as Proposal AA) proposed an amendment of Art. 60.14 to read: "60.14. Names or epithets indicated as abbreviated are to be expanded in conformity with nomenclatural tradition (see also Art. 23 *Ex. 23 and Rec. 60C.4(d))." McNeill *et al.* commented on this change:

> It might be thought that this is already covered by the provision in Art. 60.13 by which a full stop (period) in an epithet is treated as an error to be corrected by expansion or, when nomenclatural tradition does not support expansion (Art. 60.14), deletion of the full stop. However, as other indications of abbreviation are possible and as this provision provides a link to a voted example, we are not proposing a merger with Art. 60.13 ...

The Rapporteurs commented (Turland & Wiersema 2024: 390): "Prop. AA would restrict the application of Art. 60.14 to names or epithets that are themselves indicated as abbreviated, e.g. by use of a full stop (period) or other symbol, not any name or epithet that might be formed from one or more abbreviations ..." This proposal was accepted at the nomenclatural session at Madrid, 2024 (Turland *et al.* 2024: preliminary p. 12). In the cases where Forsskål eponyms have been abbreviated in the original publications, the abbreviation has always been indicated with a full stop in the epithet. The important difficulty with expansion of abbreviated epithets we have met for this work has been to decide how the epithet should be expanded, chiefly if it should be expanded in a substantival or adjectival form, as both forms have occasionally been used by later authors. We have in these cases used the expanded version in *Index Kewensis* (Jackson 1895a,

1895b) as a guide, particularly when the abbreviated eponym has not been much used after its initial publication.

Rijckevorsel (2023: 459, Proposal 228, by the Rapporteurs referred to as Proposal E) proposed to amend the second sentence of Art. 60.7 as follows: "60.7. Diacritical signs are not used in scientific names. When such signs appear in the spelling of a name at valid publication, the signs are to be suppressed with the necessary transcription of the letters so modified; [...]." The omitted text in square brackets is unchanged from the present ICN, including the words 'å becomes ao'. Rijckevorsel (2023: 459, Proposal 229, by the Rapporteurs referred to as Proposal F) proposed to rephrase the second sentence of Art. 60.7, so that it reads:

> 60.7.[...] When such a sign appears in the original spelling of a name, this sign is to be suppressed without transcription (é, è, ê become e; ñ becomes n), except a Germanic ä, ö, or ü, which is to be transcribed, to become, respectively, ae, oe, and ue; or the Scandinavian å or ø, which is also to be transcribed, to become, respectively, ao and oe. Rijckevorsel commented: It would establish that, once a name has been validly published, actual transcription (adding a new letter) is limited to just these five cases.

The Rapporteurs commented on these two proposals (Turland & Wiersema 2024: 389):

> Prop. E would offer an improved wording for the part of Art. 60.7 that concerns the non-use of diacritical signs in scientific names. Prop. F is an alternative to Prop. E, ... Proposal F would make it clearer that 'transcription' entails adding an extra letter, such as in ä becoming ae, but not in é becoming e, whereas the wording 'necessary transcription', followed by 'for example' as in the current Art. 60.7 only implies that transcription does occur in some specified cases but not in all cases and leaves open what exactly these are. Prop. F would specify that certain letters with diacritical signs are to be transcribed only when their usage is Germanic or Scandinavian, which mean that such signs in other languages would not be so transcribed, for example, that Turkish ö and ü will become o and u and not be transcribed as oe and ue.

The first of these two alternative proposals, Prop. E, was accepted by the nomenclature session in Madrid, 2024 (Turland *et al.* 2024), and 'å' is therefore also in the future to be transcribed to 'ao'. We find that Proposal F would have added more clarity to the ICN, but also make the text more complicated.

Rijckevorsel (2023: 459, Proposal 231, by the Rapporteurs referred to as Proposal H) proposed to add a new Example to Art. 60.7: "Ex. n. The letter å to become ao. *Forsskålea* (dedicated to Peter Forsskål) is to be cited as *Forsskaolea* L. (*Opobalsamum*: 17, 1764), not as *Forskohlea*, as done by Linnaeus (*Mant. Pl.*: 72, 1767), or as *Forskalea*, as done by Jussieu (*Gen. Pl.*: 403, 1789)." The Rapporteurs commented on this (Turland & Wiersema 2024: 389) that Prop. H would be automatically referred to the Editorial Committee, and this was what was confirmed at the nomenclatural session in Madrid, 2024. We hope that this example will be included in the Madrid Code.

Unlike our doubts about the interpretation of the orthography rules in the ICN sometimes expressed above, there were few cases where we have been in doubt about the validity and legitimacy or not of the eponyms.

2
Material and methods

To identify the Forsskål eponyms we have made use of all the databases cited in the chapter *Orthographic variants of eponyms …* (IPNI continuously updated; POWO continuously updated; WFO continuously updated; GBIF continuously updated; TNRS continuously updated; WoRMS continuously updated) when we searched for epithets beginning with 'Forsk', 'Forssk', 'forsk', and 'forssk', and a total of 134 records of Forsskål's plant eponyms were found. However, based on a critical scrutiny of this list and by removing errors and duplicate names, of which some were simple duplications, some represented varying orthography of the original epithet in new combinations, and some represented repetition of species names with Forsskål eponym epithets in combination with different infraspecific epithets. So far, no Forsskål eponym has been used as an infraspecific name.

A thus 'cleaned' list contained 73 Forsskål eponyms and these have been studied in detail. All literature with these 73 eponyms cited in the databases was consulted to determine the original spelling of the name. We also consulted the information in *The Plants of Pehr Forsskål's Flora Aegyptiaco-Arabica* (Hepper & Friis 1994) to see if the eponyms had been recorded there and, if so, how they had been treated. Entries in IPNI (continuously updated) and POWO (continuously updated) that could be seen in 2024 are not discussed further here because of the note with Forsskål's name as an author in IPNI (continuously updated), stating that the database will be updated once this work is published: "See Friis & Thulin in Taxon, 33(4): 668–672 (1984) re spelling of this surname. Note: In IPNI we have refrained from changing related epithets (other than transcribing diacritical signs), pending a more comprehensive review in the near future."

In some cases original herbarium material has been investigated in the herbaria BM, C (the separate Forsskål herbarium), G, K, LINN, LINN-HS, P, S, and UPS (acronyms following Thiers continuously updated), mainly by the study of the images in the database Global Plants (continuously updated), which includes high-resolution images of specimens, or through direct contact with the herbaria and their curators.

When deciding on the currently used names for the Forsskål eponyms, we have again used Hepper & Friis (1994), but also consulted the recent floras of the region, including the *Flora of Egypt* (Boulos 1999, 2000, 2002, 2005), *The Plants of Sudan and South Sudan* (Darbyshire *et al.* 2015), the *Flora of Ethiopia and Eritrea* (Edwards *et al.* 1995, 1997, 2000; Hedberg *et al.* 1989, 1995, 2003, 2004, 2006, 2009a, 2009b), the *Flora of Somalia* (Thulin 1993, 1995, 1999, 2006), and *A handbook of the Yemen Flora* (Wood 1997). We have omitted detailed referencing to these works under the treatment of the eponyms.

In the introductory chapters, where references to the same works often appear in several separate places, we have used the author-date reference-system to books, journals with authors (or database-abbreviations), and a year of publication, and then listed the works in the list of *Literature cited* in this work. For the following taxonomic part, where references generally only occur once or a few times (except for Forsskål's *Flora Aegyptiaco-Arabica* and a few others, mainly Vahl (1790, 1791, 1794), which works it is necessary to cite frequently) we have used the same reference system as for nomenclatural discussions in the journal *Taxon*, which is an author name or a string of author names, abbreviated name for books or journals, sometimes edition and volume, and page and year. It is also a convention in *Taxon* and elsewhere to have no space between letters in the initials and the abbreviated name when authors of plant names are indicated; the same applies to 'f.' when this indicates 'filius' or 'filia' [son or daughter of], except when the author's surname is not abbreviated, as, for example, is the case with Forsyth f.

In this work, we have cited abbreviated names of the authors of plant names as used in IPNI (continuously updatd). To find full names of books and journals in IPNI, it may be useful to search for the precise abbreviation as indicated here, and then continue search with the full name. Abbreviated book titles are from the online version of TL-2 (2024), and abbreviated titles of journals from the online version of BPH (continuously updated), but we have as far as possible given the full titles in the list of references.

The *International Code of Nomenclature for Algae, Fungi, and Plants* (*Shenzhen Code*; Turland *et al.* 2018) has been the basis for our nomenclatural decisions in this work, apart from the few relevant amendments that have been made at the XX International Botanical Congress in Madrid, July 2024. These amendments are in force as rules from their adoption at the closing General Assembly in Madrid on the 19th of July, 2024 (own notes by Friis; Turland *et al.* 2024). We read all proposals carefully, voted in the mail vote before the congress, and followed the live streaming of the proceedings of the Nomenclature Session of the Congress (Knapp *et al.* 2024), taking notes on the voting on all the amendments to the Code to see which ones might affect this work.

3
Plant eponyms honouring Forsskål: orthography, nomenclatural status and name in current use for each species

The eponyms are arranged according to their position in families recognized in APG IV (2016), and the families are arranged alphabetically. Within the families, the eponyms are listed alphabetically according to the spelling of the name used in the original publication, which is presented at the beginning of each entry in bold italics. Necessary and recommended corrections according to the ICN are indicated. We discuss whether the name is validly published and legitimate or not. Under each entry a summary states the Correct orthography in italics, the nomenclatural status, and the currently used name for the taxon in bold italics.

Acanthaceae

Justicia Forskoklii Zoëga in L., *Mant. Pl. Altera*: 173, 1771.
This designation has sometimes been taken as a name proposed by Zoëga and published by Linnaeus, for example by Christensen (1922: 10, where it is cited as *J.[usticia] Forskolei* Zoëga in L., *Mant.* II: 173, 1771). However, Linnaeus in *Mant. Pl. Altera*: 173, 1771, states that three descriptions of three other finds of this shrub [*Justicia fastuosa* L.] had been sent to him by Zoëga: "Descriptiones tres alias hujus fruticis addere lubet a D. Zoega missas" [as *missos*]. Johan Zoëga was a student of Martin Vahl in Copenhagen. The texts associated with the three descriptions provided by Zoëga are, for the first: * *Koenigii e frutice lecto in Insulæ Johannæ sylvis*. Fruticulus. 'Insula Johanna' is one of the Comoro Islands. For the second: * *Forskoklii – e Gemen, prope montem Melhara* [Melhan]: *fruticulus frequens. Gemen* is a typographical error for Yemen, followed by more detailed information about the locality. For the third: * Zoegæ *ex speciminibus missis facta* [factual observations from the specimens sent by Zoega]'. As we interpret the ICN, Art. 36.1, "A name is not validly published when it is not accepted by its author in the original publication …", Linnaeus has, as indicated by the wording and typography, not accepted *Koenigii* and *Forskoklii* as new names. These words only indicate that Zoëga's two first descriptions were of plants collected by König and Forsskål. They are therefore not validly published epithets. In the protologue of *Justicia Forskålei* Vahl, *Symb. Bot.* 1: 2, 1790 (see below), Vahl states that the description of that plant in *Mant. Pl. Altera*, p. [173–]174 is by Zoëga, and that it belongs to his own species, *Justicia forskaolii* [as *Forskålei*] Vahl, not to *J. fastuosa* L.
Correct orthography: No correction of *Justicia Forskoklii* Zoëga in L. is needed, as this is a designation with no nomenclatural standing.
Nomenclatural status: A designation, not a validly published name.
Currently used name for the taxon: ***Hypoestes forskaolii*** (Vahl) R.Br.

Justicia Forskålei Vahl, *Symb. Bot.* 1: 2, 1790.
According to ICN, the eponym must be corrected to *Justicia forskaolii* Vahl. – Forsskål's name, *Justicia paniculata* Forssk., *Fl. Aegypt.-Arab.*: CII, No. 20; 4, *Cent.* I, No. 4, 1775, was based on a plant from Yemen. When Forsskål's *Fl. Aegypt.-Arab.* was published in 1775, 12 years after Forsskål's death, a valid and legitimate homonym had been published for a different species, *Justicia paniculata* Burm.f., *Fl. Indica*: 9, 1768, which made *Justicia paniculata* Forssk. an illegitimate later homonym. In the protologue of his eponym, Vahl cited *J. paniculata* Forssk. in synonymy, but because that name is illegitimate, Vahl's name is a legitimate replacement name. Brown, in R.Br., *Prodr. Fl. Nov. Holland.*: 474, 1810, suggested that Vahl's

J. forskaolii should be transferred to the new genus *Hypoestes* Sol. ex R.Br. (1810), but did not make the necessary new combination. Brown later transferred *J. forskaolii* Vahl to *Hypoestes* as *H. forskaolii* [as *Forskalii*] (Vahl) R.Br., in Salt, *Voy. Abyss.*, Appendix 4: lxiii, 1814 [Sep. 1814], citing *Justicia Forskalii* Willd., *Sp. Pl.* 1: 79, 1797 [as *Forskaelei*], as the basionym. There is a later orthographic variant of Vahl's name, *Justicia forskohlii* R.Br. ex Steud., *Nomencl. Bot.*, ed. 2, 1: 838, 1840, but the authority, 'R.Br. ex Steud.' is not correct. Steudel cited *Hypoestes Forskohlei* [without authority] in synonymy, which is an orthographic variant of *H. forskaolii* (Vahl) R.Br. Mabberley (1978: 64) noted the transfer in Salt's travel account but identified the species with *Hypoestes verticillaris* (L.f.) Roem. & Schult., *Syst. Veg.*, ed. 15[bis], 1: 140, 1817; see also Mabberley & Moore (2022: 568).
Correct orthography: *Justicia forskaolii* Vahl
Nomenclatural status: Legitimate.
Currently used name for the taxon: ***Hypoestes forskaolii*** (Vahl) R.Br.

Ruellia forsskaolii Thulin, *Symb. Bot. Upsal.* 38: 142, 2016.
According to ICN, no correction of the orthography of the eponym is necessary. – Based on material from Yemen, Forsskål described this species and named it *Camellia grandiflora* Forssk., *Fl. Aegypt.-Arab.*: CXVII, No. 427; 126, *Cent.* IV, No. 99, 1775, a legitimate, but taxonomically wrong name, as the genus *Camellia* is unrelated to this species in the Acanthaceae. Vahl renamed it *Ruellia longiflora* Vahl, in *Symb. Bot.* 1: 45, Tab. 15, 1790, stating that the placement in *Camellia* was an *error editoris*, an error of the editor. However, there was no published name like *Ruellia grandiflora* to prevent a new combination based on Forsskål's name, and *R. longiflora* Vahl is a superfluous, illegitimate renaming according to ICN, Art. 52.1. Later, Persoon proposed a new combination based on Forsskål's epithet, *R. grandiflora* (Forssk.) Pers., *Syn. Pl.* 2: 175, 1806, but by then a legitimate *R. grandiflora* Poir., in Lam., *Encycl. Meth. Bot.* 6(1): 340, 1804, had been published for a plant from America, and the name *R. grandiflora* (Forssk.) Pers. was therefore an illegitimate later homonym. Nees proposed the name *Dipteracanthus longiflorus* Nees, in A.DC., *Prodr.* 11: 117, 1847, citing in synonymy '*Camellia longiflora* Forssk.' (a name which Forsskål did not publish) and the illegitimate *Ruellia longiflora* Vahl. Nees's name is a superfluous, illegitimate renaming because there was no name preventing the use of the epithet from Forsskål's *C. grandiflora* in *Dipteracanthus*. Oersted proposed a new genus, *Solaenacanthus* Oerst., in *Vidensk. Meddel. Naturhist. Foren. Kjøbenhavn* 1854: 122, 1855, and cited *Dipteracanthus longiflorus* as belonging to this new genus, but he did not formally make the combination. Müller validated the combination in *Solaenacanthus*, *S. longiflorus* (Nees) Muell.Berol., in Walpers' *Ann. Bot. Syst.* 5: 639, 1858, ascribing the combination to Oersted, but again these names are illegitimate. Blatter proposed *Ruellia grandiflora* (Forssk.) Blatt., in *Rec. Bot. Surv. India* 8: 353, 1921, but that is an isonym of *R. grandiflora* (Forssk.) Pers. After the publication of Wood, *Handb. Yemen Flora*: 270, 1997, where *Ruellia grandiflora* (Forssk.) Blatt. was accepted, Thulin (2016) concluded that a new epithet was necessary and published the eponym *Ruellia forsskaolii* Thulin.
Correct orthography: *Ruellia forsskaolii* Thulin
Nomenclatural status: Legitimate.
Currently used name for the taxon: ***Ruellia forsskaolii*** Thulin

Aizoaceae

Mesembryanthemum Forskahlei Hochst. ex Boiss., *Fl. Orient.* 2: 765, 1872–1873 [Dec. 1872 or Jan. 1873].
According to ICN, the orthography of the eponym must be corrected to *Mesembryanthemum forskahlii* Hochst. ex Boiss. – In the protologue of his eponym, Boissier cited in synonymy *Mesembryanthemum geniculiflorum* Forssk., *Fl. Aegypt.-Arab.*: LXVII, No. 273; 98, *Cent.* IV, No. 1, 1775, described from Egypt, but there was already a *M. geniculiflorum* L., *Sp. Pl.* 1: 481, 1753, described from South Africa. The Linnaean species is restricted to Namibia and South Africa. We conclude that Forsskål's name is misidentification, *M. geniculiflorum sensu* Forssk., non L. Boissier's change of epithet is therefore legitimate. Brown made a new

combination based on this, *Opophytum forskahlii* [as *Forskahlii*] (Hochst. ex Boiss.) N.E.Br., in *Gard. Chron.*, ser. 3, 84: 253, 1928. Boissier's name is either accepted as the correct name for a distinct species restricted to the dry parts of Africa north of the Equator or considered a taxonomic synonym of *M. cryptanthum* Hook.f., in *Hooker's Icon. Pl.* 11: tab. 1034, 1868, which was described from St. Helena and occurs in the dry parts of both Northern and Southern Africa. We have not seen a formal synonymy, but *M. forskahlii* is mostly placed as synonym of *M. cryptanthum* in herbaria and databases.
Correct orthography: *Mesembryanthemum forskahlii* Hochst. ex Boiss.
Nomenclatural status: Legitimate.
Currently used name for the taxon: ***Mesembryanthemum cryptanthum*** Hook.f.

Amaranthaceae (including Chenopodiaceae in Hepper & Friis 1994)

Allaganthera Forskâlii Mart., *Hort. Erlang.*: 69, 1814.
The letter 'â' in the original publication must be a typographic error for 'å,' the Scandinavian 'å' not having been available in the typesetter's case.
According to ICN, the orthography of the eponym must be corrected to *Allaganthera forskaolii* Mart. – In the protologue of his eponym, Martius cited in synonymy the generic name *Alternanthera* Forssk., *Fl. Aegypt.-Arab.*: 28, *Cent.* I, No. 100, 1775, on which page in *Fl. Aegypt.-Arab.* there is a description of a new genus without specific epithet, based on a plant from Egypt. However, on p. LIX, No. 17, there is an abbreviated species name provided by Forsskål, *Alternanthera acyranth.,* which, in agreement with nomenclatural tradition according to ICN, Art. 60.14, is to be expanded, and, as it is indicated as abbreviated by a full stop, also according to the proposed amendment by McNeill *et al.* (see above). According to *Index Kewensis* (Jackson 1895a: 95), this is to be expanded to *A. achyranthes* Forssk. Brown (R.Br., *Prodr. Fl. Nov. Holland.*: 417, 1810), suggested that an abbreviated name, *Ill. Achyrantha,* belonged to the genus *Alternanthera* Forssk. Sweet took up this formally with reference to Brown's *Prodromus,* as *A. achyrantha* Sweet, *Hort. Suburb. Londin.*: 48, 1818. Brown's *Ill. Achyrantha* must refer to *Illecebrum achyrantha* L., *Sp. Pl.*, ed. 2, 1: 299, 1762, which is a superfluous, illegitimate renaming of *Achyranthes repens* L., *Sp. Pl.* 1: 205, 1753. However, Forsskål made no reference to the earlier *Illecebrum Achyrantha* L., and the name *Alternanthera achyranthes* Forssk. is therefore validly published and legitimate. Martius also cited *Gomphrena sessilis* L., *Sp. Pl.* 1: 225, 1753, and *Illecebrum sessile* (L.) L., *Sp. Pl.*, ed. 2, 1: 300, 1762, as synonyms of his new name, for which reason the oldest epithet from *G. sessilis* L. should have been used, and Martius' name is therefore illegitimate. The currently accepted name is *Alternanthera sessilis* (L.) DC., *Cat. Pl. Horti Monsp.*: 77, 1813.
Correct orthography: *Allaganthera forskaolii* Mart.
Nomenclatural status: Illegitimate.
Currently used name for the taxon: ***Alternanthera sessilis*** (L.) DC.

Digera Forskaoli Blume [as ***DIGERA FORSKAOLI***], *Bijdr. Fl. Ned. Ind.* 11: 542, 1826 [24 Jan. 1826].
According to ICN, the eponym must be written *Digera forskaolii* Blume. – In the protologue, Blume made no further reference to Forsskål than citing him as the author of the generic name *Digera* Forssk. However, Forsskål published both the generic name *Digera* and the species name, *D. arvensis* Forssk., *Fl. Aegypt.-Arab.*: CVI, No. 141; 65, *Cent.* III, No. 7, 1775, based on a plant from Yemen, and both the genus name and the species name are legitimate. In the absence of a direct or indirect reference to Forsskål's species, Blume's eponym must have been given to a new species from Java, the name of which honours Forsskål without relationship to his material or names, and it is therefore legitimate. *Digera forskaolii* Blume is now a taxonomic synonym of *D. muricata* (L.) Mart., *Nova Acta Phys.-Med. Acad. Caes. Leop.-Carol. Nat. Cur.* 13: 285 (1826).
Correct orthography: *Digera forskaolii* Blume
Nomenclatural status: Legitimate.
Currently used name for the taxon: ***Digera muricata*** (L.) Mart.

Salsola Forskahlii Forsyth f., *Bot. Nomencl.*: Column 169, 1794.
And
Salsola Forskalii Schweinf., in *Bull. Herb. Boissier* 4, Append. 2: 160, 1896.
According to ICN, the orthography of Forsyth f.'s eponym should be written *Salsola forskahlii* Forsyth f., and Schweinfurth's eponym as *S. forskalii* Schweinf. – The two names are based on different types and represent different Latinizations. In the protologue of his new eponym, Forsyth cited in synonymy the name '*Suaeda anonyma 503*'. Forsskål's genus name *Suaeda* is published with description of seven species in *Fl. Aegypt.-Arab.*: LXIV, No. 186–187, CIX No. 22; 69–70, *Cent.* III, No. 15–18, 1775, but since there is no separate generic description of the new genus *Suaeda*, none of the new names for species are validly published, and an unnamed species on p. 69, *Cent.* III, No. 17. *Suaeda; foliis oblongis; interdum linearibus, vix pollicaribus. Arab. Hömmäm* [Suaeda with oblong, sometimes linear leaves, scarcely the length of the first joint of thumb. In Arabic Hömmam] has no counterpart in the *Florulae*, although the vernacular name Hömmam is cited for *Suaeda vera* in both *Flora Aegyptiaca* and *Flora Arabiœ Felicis*. The genus name *Suaeda* only became validly published in J.F.Gmel., *Suaeda* Forssk. ex J.F.Gmel., *Onomat. Bot. Compl.* 8: Column 797, 1776, nom. cons., with a description of the genus and descriptions of Forsskål's seven species in two columns in the German language, including *Suaeda anonyma* J.F.Gmel., *Onomat. Bot. Compl.* 8: Column 798, 1776. Gmelin again used the name *S. anonyma* J.F.Gmel. in his *Syst. Nat.*, ed. 13bis. 2(1): 503, 1791, hence Forsyth's reference to '503'. Schweinfurth (1912: 155) was the first to identify the unnamed species of *Suaeda* with *S. vera* Forssk. ex J.F.Gmel., an identification upheld by Christensen (1922: 22) and Hepper & Friis (1994: 111). Gmelin's epithet *anonyma* is permitted under the ICN and should have been accepted by Forsyth f. in 1794. Forsyth's name therefore represents a superfluous, illegitimate change of epithet from *Suaeda anonyma* J.F.Gmel., a synonym of *S. vera* Forssk. ex J.F.Gmel. – When publishing *Salsola forskalii* Schweinf. without mentioning *Salsola forskahlii* Forsyth f., Schweinfurth did not base his eponym on material collected by Forsskål or on names published by him but stated that it was a new species based on material he had himself collected in Eritrea, south-eastern Egypt, Yemen and Aden (see about Schweinfurth's expedition in Christensen, 1918: 86), and that he had chosen to name it in honour of Forsskål. Schweinfurth compared his new species with *Salsola vermiculata* L., *Sp. Pl.* 1: 223, 1753, but argued in favour of it being a distinct species. Schweinfurth's syntypes (*Schweinfurth 95*, Eritrea; *Schweinfurth 138, 158*, Yemen) could not be traced, and the name may need to be neotypified. Friis & Gilbert have both in Thulin (ed.), *Fl. Somalia* 1: 138, 1993, and in Edwards *et al.* (eds.), *Fl. Ethiopia & Eritrea* 2(1): 296, 2000, identified *Salsola forskalii* Schweinf. with *S. spinescens* Moq. The name *Salsola forskalii* Schweinf. is orthographically similar to the earlier, but illegitimate *S. forskahlii* Forsyth f., and in spite of the 'h' in the name proposed by Forsyth, his and Schweinfurth's names may be confused because they apply to related taxa. We are of the opinion that they are to be treated as homonyms in agreement with Art. 53.2 of the ICN, and Schweinfurth's name is therefore an illegitimate later homonym (ICN Art. 53.2). *Salsola forskalii* Schweinf. is a taxonomic synonym of *Caroxylon spinescens* (Moq.) Akhani & Roalson, *Int. J. Pl. Sci.* 168(6): 948 (2007), based on *Salsola spinescens* Moq., in A.DC., *Prodr.* 13(2): 179, 1849.
For ***Salsola Forskahlii*** Forsyth f.
Correct orthography: *Salsola forskahlii* Forsyth f.
Nomenclatural status: Illegitimate.
Currently used name for the taxon: ***Suaeda vera*** Forssk. ex J.F.Gmel.
For ***Salsola Forskalii*** Schweinf.
Correct orthography: *Salsola forskalii* Schweinf.
Nomenclatural status: Illegitimate.
Currently used name for the taxon: ***Salsola spinescens*** Moq. or ***Caroxylon spinescens*** (Moq.) Akhani & Roalson

Annonaceae

Annona Forskahlii DC., *Syst. Nat.* 1: 472, 1817 [1818 publ. 1–15 Nov. 1817].
According to ICN, the eponym should be written *Annona forskahlii* DC. – In the protologue of his eponym, Candolle stated that he considered the material of *Annona* from Egypt sufficiently distinct from the earlier *Annona squamosa* L., *Sp. Pl.* 1: 537, 1753, to represent a separate species to be called *A. forskahlii* DC., and he cited in synonymy Forsskål's name *A. glabra* Forssk., based on a plant from Egypt. When *Annona glabra* Forssk., *Fl. Aegypt.-Arab.*: LXVIII, No. 286, CXIV, No. 347; 102, *Cent.* IV, No. 16, 1775, was published, there was already an *A. glabra* L., *Sp. Pl.* 1: 537, 1753, used for a species cultivated in America and West Africa. Since Forsskål had access to the description of *A. glabra* L. on the expedition, we must conclude that Forsskål's name represents a misidentification, *A. glabra sensu* Forssk., non L. Candolle's name is therefore legitimate, but it is now considered a taxonomic synonym of *A. squamosa* L., *Sp. Pl.* 1: 537, 1753.
Correct orthography: *Annona forskahlii* DC.
Nomenclatural status: Legitimate.
Currently used name for the taxon: ***Annona squamosa*** L.

Apocynaceae (including Asclepiadaceae in Hepper & Friis 1994)

Asclepias Forsk. Roem. & Schult., *Syst. Veg.*, ed. 15[bis], 6: 85. 1820 [Aug.–Dec. 1820].
According to ICN (including the accepted proposal at the Madrid nomenclatural session, McNeill *et al.* 2022), the eponym is clearly indicated as abbreviated with a full stop and is to be expanded to a full name in conformity with nomenclatural tradition. Traditionally, the abbreviated epithet *forsk.* has been expanded as *Asclepias forskohlei* Roem. & Schult., for example by Jackson (1895a: 208), which, corrected as *A. forskohlii* Roem. & Schult., we accept as representing the nomenclatural tradition. – In the protologue of their eponym, Roemer & Schultes cited in synonymy *Ascl. nivea* Forssk., *Fl. Aegypt.-Arab.*: 51, [*Cent.* II,] No. 73, 1775, and *Ascl. nivalis* J.F.Gmel., *Syst. Nat.*, ed. 13bis, 2(1): 444, 1791. *Asclepias nivea* Forssk. was based on a plant from Yemen, but there was already an *A. nivea* L., *Sp. Pl.* 1: 215, 1753, described from America. We conclude that Forsskål's name is a misidentification, *A. nivea sensu* Forssk., non L. In his protologue of *A. nivalis*, Gmelin referred to Forssk., *Fl. Aegypt.-Arab*, p. 51, 1775, and thereby proposed a new, legitimate name for Forsskål's *A. nivea*. However, Roemer & Schultes also cited Gmelin's name in synonymy, and their new eponym is therefore a superfluous, illegitimate renaming of *A. nivalis* J.F.Gmel. The currently accepted name is *Pentatropis nivalis* (J.F.Gmel.) D.V.Field & J.R.I.Wood, in *Kew Bull.* 38(2): 215, 1983.
Correct orthography: *Asclepias forskohlii* Roem. & Schult.
Nomenclatural status: Illegitimate.
Currently used name for the taxon: ***Pentatropis nivalis*** (J.F.Gmel.) D.V.Field & J.R.I.Wood

Desmidorchis Forskalii Decne., in *Ann. Sci. Nat., Bot., Sér.* 2, 9: 265, 1838.
And
Boucerosia Forskalii Decne., in DC., *Prodr.* 8: 648, 1844 [mid Mar. 1844].
And
Caralluma forsskaolii Plowes, in *Asklepios* 50: 16, 1990.
According to ICN, Decaisne's eponyms should be written *Desmidorchis forskalii* Decne. and *Boucerosia forskalii* Decne. – In the protologue of *D. forskalii*, Decaisne cited *Stapelia quadrangula* Forssk., *Fl. Aegypt.-Arab.*: CVIII, No. 190; 52, *Cent.* II, No. 76, 1775, & *Icones*: Tab. 6, 1776, described from Yemen, and Forsskål's name is legitimate. There was no epithet in *Desmidorchis* preventing the transfer of Forsskål's epithet, and Decaisne's eponym is therefore a superfluous, illegitimate renaming. *Boucerosia forskalii* Decne. and *Caralluma forsskaolii* Plowes are also illegitimate renamings of *Stapelia quadrangula* Forssk. Currently used names are: *Caralluma quadrangula* (Forssk.) N.E. Br, *Gard. Chron.*, ser. 3, 12: 278, 1892; *Monolluma quadrangula* (Forssk.) Plowes., in *Haseltonia* 3: 65 (1995);

or *Ceropegia quadrangula* (Forssk.) Bruyns, in *S. African J. Bot.* 112: 412 (2017).
Correct orthography: *Desmidorchis forskalii* Decne.; *Boucerosia forskalii* Decne.; *Caralluma forsskaolii* Plowes
Nomenclatural status: Illegitimate.
Currently used name for the taxon: ***Caralluma quadrangula*** (Forssk.) N.E.Br., ***Monolluma quadrangula*** (Forssk.) Plowes, or ***Ceropegia quadrangula*** (Forssk.) Bruyns.

Sarcostemma Forskålianum Schult., *Syst. Veg.*, ed. 15bis, 6: 117, 1820 [Aug.–Dec. 1820].
According to ICN, the orthography of the eponym must be corrected to *Sarcostemma forskaolianum* Schult. – In the protologue of his eponym, Schultes cited in synonymy Forsskål's name *Asclepias aphylla* Forssk., *Fl. Aegypt.-Arab.*: CVIII, No. 187; 50, *Cent.* II, No. 69, 1775, based on a plant from Yemen. Forsskål's name is legitimate and the epithet *aphylla* should have been adopted by Schultes, but on p. 116 in the same publication, Schultes also made the new combination *S. aphyllum* R.Br. ex Schult. Brown, in *Mem. Wern. Nat. Hist. Soc.* 1: 51, 1811, did not make the combination *Sarcostemma aphyllum* but only stated that in the new genus *Sarcostemma* "the probable species are, *Asclepias aphylla*, Thunb. Prod., ..." and a number of other species, among which Brown did not mention *Asclepias aphylla* Forssk. In his protologue of *S. aphyllum*, Schultes cited *A. aphylla* Thunb., *Prodr. Pl. Cap.* 1: 47, 1794, based on a plant from South Africa, and this name is also indirectly cited by his reference to 'Willd. Spec I. p. 1262.' The citations in Schultes's protologue of *Sarcostemma aphyllum* refer to various components in the protologue of Thunberg's *A. aphylla*, including the original material of Thunberg's collection from the Cape, "Karro trans Hartequas Kloof prope Hexrivier", but not to Forsskål's plant from Yemen. Hepper & Friis (1994: 83) found no original material of Forsskål's plant, which was collected on Jebel Melhan in Yemen. Forsskål's and Thunberg's two identical names are obviously based on different original material, and Thunberg's name, published 31 years after Forsskål's death, is a later homonym of Forsskål's. In his protologue of *S. aphyllum*, Schultes provided a new description of Thunberg's plant and referred to Thunberg's collection from South Africa. According to ICN, Art. 11.4. Note 2 and Ex. 18, the name *S. aphyllum* Schult. precludes any simultaneous (or later) homonymous combination with the same final epithet that might require (or had been requiring) transfer to the same genus, irrespective of the priority of Forsskål's name. According to this, the name *S. aphyllum* is to be ascribed to Schultes alone, it is legitimate and takes priority from 1820. Don renamed *S. aphyllum* Schult. as *S. thunbergii* G.Don, *Gen. Hist.* 4: 156, 1837, citing Thunberg's and Schultes's names in synonymy. Therefore, G.Don's name is a superfluous renaming of *S. aphyllum* Schult. The taxon was reduced first to a subspecies of *S. viminale* (L.) R.Br., *Prodr. Fl. Nov. Holland.*: 464, 1810, as *S. viminale* subsp. *thunbergii* (G.Don) Liede & Meve, *Bot. J. Linn. Soc.* 112(1): 10, 1993, and later to a subspecies of *Cynanchum viminale* (L.) L., as '*C. viminale* subsp. *thunbergii* (G.Don) Liede & Meve,' in *Kew Bull.* 67(4): 754, 2012 [6 Sep. 2012; epublished]. Liede & Meve's names for this subspecies are legitimate but due to the fact that Thunberg's name is illegitimate, they are to be ascribed to Liede & Meve alone. Returning to Schultes's Forsskål eponym, Meve & Liede made the new combination *Cynanchum forskaolianum* (Schult.) Meve & Liede, in *Kew Bull.* 67: 752, 2012; they based this correctly on the legitimate *S. forskaolianum* Schult. and neotypified the name, because the type of Forsskål's *Asclepias aphylla* has not been retrieved.
Correct orthography: *Sarcostemma forskaolianum* Schult.
Nomenclatural status: Legitimate.
Currently used name for the taxon: ***Cynanchum forskaolianum*** (Schult.) Meve & Liede

Daemia [Schultes's misspelling for ***Doemia***]? ***Forskåli*** Schult., *Syst. Veg.*, ed. 15bis, 6: 113, 1820 [Aug.–Dec. 1820].
According to ICN, the orthography of the eponym must be corrected to *Doemia forskaolii* Schult. (Fig. 9). – In the protologue of his eponym, Schultes cited *Asclepias daemia* Forssk., *Fl. Aegypt.-Arab.*: 51,

Cent. II, No. 71, 1775, based on a plant from Yemen, which is a validly published and legitimate name, although there is no species with that name among the other species of *Asclepias* in the floristic list on p. CVIII of *Flora Aegyptiaco-Arabica*. *Daemia forskaolii* Schult. is a legitimate renaming because a transfer of *Asclepias daemia* Forssk. to the genus *Daemia,* as spelt by Schultes, would have resulted in a tautonym (not permitted under the ICN, Art. 23.4). However, in the original publication, Robert Brown wrote the genus name as *Doemia* R.Br., in *Asclepiadaceae*: 39, 1810, and *Mem. Wern. Nat. Hist. Soc.* 1: 50, 1811, with ligature between 'o' and 'e'. Although Greuter *et al.* (1993: 315) recorded the spelling *Daemia* as the name in current use, it is now accepted that the correct spelling is *Doemia* (https://naturalhistory2.si.edu/botany/ing/genusSearchTextMX.cfm), which makes the combination *Doemia daemia* permissible, and hence Schultes's *Daemia forskaolii* is illegitimate. Chiovenda transferred *A. daemia* Forssk. to *Pergularia* as *P. daemia* (Forssk.) Chiov., *Res. Sci.* [Missione Stefanini-Paoli] *Somalia Ital.* 1: 115, 1916.
Correct orthography: *Doemia forskaolii* Schult.
Nomenclatural status: Illegitimate.
Currently used name for the taxon: ***Pergularia daemia*** (Forssk.) Chiov.

Kanahia Forskalii Decne., in DC., *Prodr.* 8: 537, 1844 [mid Mar. 1844].
According to ICN, the eponym should be written *Kanahia forskalii* Decne. – In the protologue of his eponym, Decaisne cited in synonymy *Asclepias laniflora* Forssk., *Fl. Aegypt.-Arab.*: CVIII, No. 180; 51, *Cent.* II, No. 72, 1775, based on a plant from Yemen. Forsskål's name is valid and legitimate, was accepted by Vahl and cited in *Symb. Bot.* 1: 23 & Tab. 7, 1790. Forsskål's epithet was transferred to *Kanahia* by Brown in the new combination *K. laniflora* (Forssk.) R.Br., in Salt, *Voy. Abyss.*, Append.: lxiv, 1814 [as *Kannahia laniflora*], see also Mabberley & Moore (2022: 504). Decaisne's eponym is therefore a superfluous, illegitimate renaming.
Correct orthography: *Kanahia forskalii* Decne.
Nomenclatural status: Illegitimate.
Currently used name for the taxon: ***Kanahia laniflora*** (Forssk.) R.Br.

Leptadenia Forskoelii G.Don, *Gen. Hist.* 4: 124, 1838 [8 Mar.–8 Apr. 1838].
And
Leptadenia Forskalii Decne., in *Ann. Sci. Nat.*, Ser. 2, 9: 269, 1838 [May or Jun. 1838].
The protologue of the former name was published without indication of original material, except that it was collected by Forsskål in Yemen; the letters 'oe' in the eponym are written together as a ligature, 'œ'. The protologue of the latter name states that it is based on a specimen from Yemen collected by Botta.

According to ICN, G.Don's eponym should be written *Leptadenia forskoelii* G.Don and that of Decaisne *L. forskalii* Decne. The dates of publication of the two eponyms are indeed close, but according to TL-2 (2024), G.Don's name must have priority. – Neither Don, nor Decaisne cited names proposed by Forsskål, but their names are commemorative of him and validly published. Bruyns *et al.*, in *Taxon* 72(1): 136, 2023, have summarised all lecto- or neotypifications of Forsskål's names now in *Leptadenia*. *Cynanchum arboreum* Forssk., *Fl. Aegypt.-Arab.*: CVIII, No. 177; 53, *Cent.* II, No. 80, 1775, now *L. arborea* (Forssk.) Schweinf., *Arab. Pfl.-Nam.*: 167, 1912, was lectotypified by Bruyns & Miller in *Edinburgh J. Bot.* 66: 99, 2009, with a Forsskål collection from Yemen, *Forsskål 271* (C barcode C10002125). *Leptadenia forskoelii* G.Don was neotypified fourteen years later by Bruyns *et al.*, in *Taxon* 72(1): 136, 2023, with the same Forsskål specimen, which makes *L. forskoelii* G.Don a homotypic synonym of *Cynanchum arboreum* Forssk. *Leptadenia forskalii* Decne. is according to Bruyns *et al.* typified with a holotype: Yemen, Hais, 1837, *Botta s.n.* (P barcode P005572932). Bruyns *et al.* state that the neotype of *L. forskoelii* G.Don is "selected among the many collections of Forsskål of this species at Copenhagen." But Hepper & Friis (1994: 82) recorded only two sheets of this species at C, *Forsskål 271* and *Forsskål 273*, and no more collections have been found since. It is not clear why Bruyns *et al.* chose to select the neotype of *L. forskoelii* G.Don to be identical with the

lectotype previously selected for *C. arboreum* Forssk. when another specimen was available. We are of the opinion that *Leptadenia forskoelii* G.Don and *L. forskalii* Decne. are so similar that they are likely to be confused, also because they are now homotypic synonyms of *L. arborea* (Forssk.) Schweinf. They are therefore to be treated as homonyms in agreement with Art. 53.2 of the ICN.
Correct orthography: *Leptadenia forskoelii* G.Don; *Leptadenia forskalii* Decne.
Nomenclatural status: *Leptadenia forskoelii* G.Don is legitimate, although a homotypic synonym of L. *arborea* (Forssk.) Schweinf. *Leptadenia forskalii* Decne. is according to us an illegitimate later homonym of *Leptadenia forskoelii* G.Don.
Currently used name for the taxon: ***Leptadenia arborea*** (Forssk.) Schweinf.

Asparagaceae (Convallariaceae in Hepper & Friis 1994)

Smilacina Forskaliana Schult. & Schult.f., *Syst. Veg.*, ed. 15[bis], 7(1): 304, 1829.
According to ICN, the orthography of the eponym should be written *Smilacina forskaliana* Schult. & Schult.f. Two new combinations, *Sansevieria forskaliana* (Schult. & Schult.f.) Hepper & J.R.I.Wood, in *Kew Bull.* 38(1): 83, 1983 [11 Aug. 1983], and *Dracaena forskaliana* (Schult. & Schult.f.) Byng & Christenh., *Global Fl.* 4: 65, 2018 [9 Feb 2018, epublished] have been based on this name. – In the protologue of their eponym, Schultes and Schultes f. cited in synonymy *Convallaria racemosa* Forssk., *Fl. Aegypt.-Arab.*: CX No. 236; 73, *Cent.* III, No. 28, 1775, based on a plant from Yemen, pointing out that it is distinct from the American *C. racemosa* L., *Sp. Pl.* 1: 315, 1753. In agreement with other cases where Forsskål's plant shares its name with an American species described by Linnaeus, we conclude that Forsskål's name, in spite of the statement by Schultes & Schultes f., is a misidentification, *C. racemosa sensu* Forssk., non L. Schultes and Schultes f. thus correctly changed the epithet, and their name is legitimate.
Correct orthography: *Smilacina forskaliana* Schult.& Schult.f.
Nomenclatural status: Legitimate.
Currently used name for the taxon: ***Sansevieria forskaliana*** (Schult. & Schult.f.) Hepper & J.R.I. Wood or ***Dracaena forskaliana*** (Schult. & Schult.f.) Byng & Christenh.

Asteraceae (Compositae in Hepper & Friis 1994)

Calendula Forskohleana hort. ex Steud., *Nomencl. Bot.*, ed. 2, 1: 256 (left column), 1840; ***Calendula Forskohliana*** Quorun.[dam] ex Steud., *Nomencl. Bot.*, ed. 2, 1: 256 (right column).
According to ICN, Steudel's eponym is to be written as *Calendula forskohliana*. – Steudel published both of his two variants in synonymy, *Calendula forskohleana* hort. as synonym of *Calendula stellata* [without authority], and *Calendula forskohliana* Quorun[dam] ('by some', meaning unknown authors) as synonym of *Calendula stellata* Cav. Both must be identified as *C. stellata* Cav., *Icon.* 1: 3 & Tab. 5, 1791, which was published later than Forsskål's *Fl. Aegypt.-Arab.* In that work, Forsskål mentioned the genus name *Calendula* three times, all as '*Calendula officin[alis* L.],' first from Istanbul (XXXIII, No. 393), then from Cairo (LXXIV, No. 470) and finally from Yemen (CXX, No. 513). In the protologue of *C. stellata* Cav., there is no direct or indirect reference to Forsskål or to *C. officinalis* L., *Sp. Pl.* 2: 921, 1753. *Calendula forskohleana* hort. ex Steud. and *Calendula forskohliana* Quorun[dam] ex Steud., are both published in synonymy of *C. stellata* Cav. (currently considered specifically distinct from *C. officinalis* L.) and not validly published.
Correct orthography: *Calendula forskohliana* hort. ex Steud.
Nomenclatural status: Not validly published.
Currently used name for the taxon: ***Calendula officinalis*** L. or ***C. stellata*** Cav.

Crepis Forskalii Babc., in *Univ. Calif. Publ. Bot.* 22: 898, 1947.
According to ICN, the eponym should be written

Crepis forskalii Babc. – In the protologue of his eponym, Babcock based the name on collections made by Schweinfurth during his expedition to Yemen in 1889–1890 (Christensen, 1918: 86) and on a specimen collected by Forsskål, presumably in Yemen, and numbered as *Forsskål 479* in B and UC, but lacking in the main Forsskål herbarium at C. The name is not cited by Hepper & Friis (1994), and not associated with a name published in Forsskål's *Fl. Aegypt.-Arab. Crepis forskalii* Babc. is a legitimate name, and according to Babcock, the species is part of a complex of forms with the currently accepted name *C. rueppellii* Sch. Bip., in *Flora* 22(1): 20, 1839.
Correct orthography: *Crepis forskalii* Babc.
Nomenclatural status: Legitimate.
Currently used name for the taxon: ***Crepis rueppellii*** Sch.Bip.

Gnaphalium Forskåhlii J.F.Gmel., *Syst. Nat.*, ed. 13bis, 2(2): 1214, 1792 [1791 publ. late Apr.–Oct. 1792]. The sign over the 'a' in the original publication is the lower part of an 'o' or an 'u' and the letter must be interpreted as an 'å,' the Scandinavian 'å' not being available in the typesetter's case (Fig. 10).
According to ICN, the orthography of the eponym must be corrected to *Gnaphalium forskaohlii* J.F.Gmel., but Räuschel, *Nomencl. Bot.*, ed. 3: 238, 1797, cited Gmelin's eponym without a transcribed 'å', thus rendering the name as '*Gnaphalium forskahlii.*' – In the protologue of his eponym, Gmelin cited in synonymy *G. fruticosum* Forssk., *Fl. Aegypt.-Arab.*: CXIX, No. 491; 218, *Cent.* VIII, No. 91, 1775, based on a plant from Yemen. This is seemingly a later homonym of the validly published and legitimate *G. fruticosum* Mill., *Gard. Dict.*, ed. 8: no. 20, 1768, which had been published after Forsskål's death but before the publication of *Flora Aegyptiaco-Arabica*. Forsskål's name has therefore been considered illegitimate and Gmelin's eponym legitimate (but see below). Later, Vatke based his *Helichrysum fruticosum* Vatke, in *Linnaea* 39(6): 491, 1875, on *G. fruticosum* Forssk. However, this traditional nomenclatural interpretation and status of '*Gnaphalium fruticosum*' Forssk. was rejected by Hilliard and Burtt, in *Notes Roy. Bot. Gard. Edinburgh* 38(1): 146, 1980. They pointed out that the apparent binomial, *Gnaphalium fruticosum* Forssk., was not published as an accepted name in *Flora Aegyptiaco-Arabica*. The typography, italics, indicated that *fruticosum* was a descriptive term and part of a phrase name, not a specific epithet (see ICN, Art. 23.6). Hilliard and Burtt are right, for the word *fruticosum* stands out in italics both on p. CXIX, No. 491, where the words *fruticosum flavum* are printed in the place where an epithet should be found, and on p. 218, *Cent.* VIII, No. 91, 1775, and flavum is in small Roman uppercase. Words in italics are used for diagnoses or phrase names in several other entries in *Gnaphalium* in *Fl. Aegypt.-Arab.* and everywhere else small Roman upper-case letters are used for species names. Moreover, Gnaphalium *fruticosum flavum*, as it is written on p. CIX No. 491, cannot be a permissible species designation according to ICN, Art. 23–6(c) (words not intended as specific epithets: "a generic name followed by two or more adjectival words in the nominative case"). Therefore, Gmelin's eponym, *G. forskaohlii,* was the first validly published name for Forsskål's species, and this made a new combination necessary, *H. forskaohlii* (J.F.Gmel.) Hilliard & B.L. Burtt, in *Notes Roy. Bot. Gard. Edinburgh* 38: 146, 1980. This taxon is sometimes considered a subspecies of *Helichrysum cymosum* (L.) D.Don, as *H. cymosum* subsp. *fruticosum* Hedb., in *Symb. Bot. Ups.* 15(1): 203, 1957, which is validly published and legitimate, being a new subspecies with a new name.
Correct orthography: *Gnaphalium forskaohlii* J.F.Gmel.; *Helichrysum forskaohlii* (J.F.Gmel.) Hilliard & B.L.Burtt
Nomenclatural status: Legitimate.
Currently used name for the taxon: ***Helichrysum forskaohlii*** (J.F.Gmel.) Hilliard & B.L.Burtt

Hieracium Forskohlei Froel., in DC., *Prodr.* 7(1): 223, 1838 [late Apr. 1838].
According to ICN, the eponym must be corrected to *Hieracium forskohlii* Froel. – In the protologue of his eponym, Froelich cited in synonymy Forsskål's name *H. hispidum* Forssk., *Fl. Aegypt.-Arab.*: XXXI, No. 342; 216, *Cent.* VII, No. 78, 1775, based on a plant from Constantinople [Istanbul]. Forsskål's name is validly

published and legitimate, and Froelich's name is therefore a superfluous, illegitimate renaming. Froelich mentioned that there was material of this species in Vahl's herbarium, but Hepper & Friis (1994: 123) found no such material, and it was not accounted for among the species of *Hieracium* in Davis, *Fl. Turkey* 5 (1975). Therefore, the identity of *Hieracium hispidum* Forssk. and *H. forskaohlii* Froel. remains unsolved.
Correct orthography: *Hieracium forskohlii* Froel.
Nomenclatural status: Illegitimate.
Currently used name for the taxon: No currently accepted name of this species is known.

Senecio forskahlii Baill., *Hist. Pl.* 8: 308, 1882 [Jan.–Feb. 1882].
According to ICN, the orthography of the eponym is correct. – In the protologue of his eponym, in footnote 4 on p. 308, Baillon cited in synonymy, but slightly incorrectly, Forsskål's name "*Kleinia? odora* Forssk., *Fl. Aegypt.-Arab.*: 146, 1775". In Forsskål's work the name is in fact written as *Cacalia odora* in both places where the name occurs, *Fl. Aegypt.-Arab.*: CXIX, No. 483, and 146, *Cent.* V, No. 66, 1775, and the name is based on a plant from Yemen. Vahl, in *Symb. Bot.* 3: 90, 1794, accepted *C. odora* and indicated Forsskål as author. A.P. de Candolle cited *C. odora* Forssk. as the basis of his new combination *Kleinia odora* (Forssk.) DC., *Prodr.* 6: 339, 1838. Forsskål's name is validly published and legitimate, and so is Candolle's new combination. Baillon's name is a superfluous, illegitimate renaming.
Correct orthography: *Senecio forskahlii* Baill.
Nomenclatural status: Illegitimate.
Currently used name for the taxon: ***Kleinia odora*** (Forssk.) DC.

Berberidaceae

Berberis Forskaliana C.K.Schneid., in *Bull. Herb. Boissier*, Ser. 2, 5: 456, 1905.
According to ICN, the eponym should be written as *Berberis forskaliana* C.K.Schneid. – In the protologue, Schneider did not cite any reference to Forsskål's work. He stated that his new species was described on material in the herbarium Barbey-Boissier (G), with a new name intended to commemorate Forsskål for his work on the flora of Yemen. The protologue is provided with a description and one collection is cited, *Schweinfurth 1682*, made by Schweinfurth on his expedition to Yemen in 1889–1890 (Christensen 1918: 86). Of this collection, we have traced duplicates at G and K. There is one mentioning of the genus *Berberis* in Forssk., *Fl. Aegypt.-Arab.*: CIX, No. 230, 1775, collected from Kurma in Yemen, and one rather poor sterile specimen is in the Forsskål collections in Copenhagen, *Forsskål 701,* but no species name is proposed on that collection. There is no connection between Forsskål's material and publication and Schneider's eponym. Wood (1997: 66) has suggested that *Berberis forskaliana* is a synonym of *Berberis holstii* Engl., *Abh. Königl. Akad. Wiss. Berlin* 1894: 69, 1894, which is widely distributed in Eastern Africa.
Correct orthography: *Berberis forskaliana* C.K. Schneid.
Nomenclatural status: Legitimate.
Currently used name for the taxon: ***Berberis forskaliana*** C.K.Schneid. or ***B. holstii*** Engl.

Boraginaceae

Echinospermum Forskohlii Lehm. in Brand, in Engl., *Pflanzenr.* IV(252), Borrag.-Borraginoid.: 152, 1931.
According to ICN, the eponym should be written *Echinospermum forskohlii* Lehm. in Brand. – However, Brand published Lehmann's eponym from a label name by 'Lehm.' (presumably, this is Johann Georg Christian Lehmann) and cited it in the synonymy of *Cynoglossum microglochin* Benth., in J.F. Royle, *Ill. Bot. Himal. Mts.*: 305, 1836. Because the eponym is published in synonymy, it is not validly published. The distribution of *Cynoglossum microglochin* does not overlap the areas through which Forsskål travelled.
Correct orthography: *Echinospermum forskohlii* Lehm. in Brand
Nomenclatural status: Not validly published.
Currently used name for the taxon: ***Cynoglossum microglochin*** Benth.

Brassicaceae (Cruciferae in Hepper & Friis 1994)

Brassica Forskåli Schult., *Observ. Bot.*: 131, No. 1012, 1809.
According to ICN, the orthography of the eponym must be corrected to *Brassica forskaolii* Schult. – Schultes's work, including this eponym, consists of proposed corrections to Willdenow's edition of *Species Plantarum*. In the protologue of his eponym, Schultes gave the reference "p. 551, No. 16. … *Ultimo loco ante secundam Categoriam ponenda*" [to be placed in the last place before the second category of *Brassica,* which is a group of species now referred to *Eruca*]. This must mean that *B. forskaolii* Schult. is intended to follow species No. 16 in Willdenow's edition of *Sp. Pl.* 3(1): 551, 1800, which is *B. teretifolia* Desf., *Fl. Atlant.*: 94, 1798. In the protologue of his eponym, Schultes provided a description, but no reference to Forsskål, no reference to locality, and no material is cited that can associate Schultes's name with names proposed by Forsskål, apart from the eponym. Both Schultes's *B. forskaolii* and the only new species of *Brassica* in Forsskål's work, *B. crassifolia* Forssk., *Fl. Aegypt.-Arab.*: LXIX, No. 323; 118, *Cent.* IV, No. 63, 1775, which was based on a plant from Egypt, have pinnate leaves with terete lobes, but the details of the descriptions of *B. crassifolia* Forssk. and that of *B. forskaolii* Schult. differ in several characters. Moreover, *Brassica crassifolia* Forssk. is recorded by Willdenow, *Sp. Pl.*, vol. 3(1): 554, 1800, and not on p. 551, to which page Schultes refers. *Index Kewensis* (Jackson 1895a: 335) recorded *B. forskaolii* Schult. as an accepted name, not a synonym, and without reference to *B. crassifolia* Forssk. We therefore consider it a validly published, legitimate name. Delile transferred *B. crassifolia* to the genus *Erucaria* as *E. crassifolia* (Forssk.) Delile, *Descr. Egypte Hist. Nat.*: 244, Tab. 34(1), 1813, but he did not comment on the taxonomic position of Schultes's *B. forskaolii.* POWO (continuously updated) has identified *B. forskaolii* Schult. with *Erucaria crassifolia* (Forssk.) Delile, but typification is needed to finally settle the identity.
Correct orthography: *Brassica forskaolii* Schult.
Nomenclatural status: Legitimate.
Currently accepted name for the taxon: ***Erucaria crassifolia*** (Forssk.) Delile

Caryophyllaceae (including Illecebraceae in Hepper & Friis 1994)

Pteranthus Forskaolii Mirb., *Hist. Nat. Pl.*, ed. 2, 10: 130, 1805.
According to ICN, the eponym should be written *Pteranthus forskaolii* Mirb. – In his treatment of the genus *Pteranthus* Forssk., Mirbel accepted it as a distinct monotypic genus and cited *Pteranthus* Forssk., *Fl. Aegypt.-Arab.*: 36, *Cent.* II, No. 17, 1775, which refers to a page in *Fl. Aegypt.-Arab.* where the genus was described without a specific epithet, based on a plant from Egypt. Mirbel therefore proposed the species name *P. forskaolii* Mirb. In synonymy of this, Mirbel cited the names *Louichea pteranthus* (L.) L'Hér., *Stirp. Nov.* 6: 135, tab. 65, 1791 [as 1785] [first published as *Camphorosma pteranthus* L., *Mant. Pl.* 1: 41, 1767, & *Syst. Nat.*, ed. 12, 2: 128, 1767, and *Louichea cervina* L'Hér., in *Louichea*: sine. pag. 1788]. The genus *Camphorosma* L., *Sp. Pl.* 1: 122, 1753, was proposed by Linnaeus with three species from the Mediterranean region, to which was added *C. pteranthus* L. in 1767, grown in the Upsala Botanical Garden from Forsskål's seeds, and *L. cervina* L'Hér. in 1788 (Iamonico *et al.* 2015). To avoid a tautonym (ICN, Art. 23.4), Mirbel could not transfer the epithet *pteranthus* to Forsskål's genus *Pteranthus*, and he had not noticed that the species name *P. dichotomus* Forssk., *Fl. Aegypt.-Arab.*: LXII, No. 100, 1775, had been published elsewhere in the *Fl. Aegypt.-Arab.* with the reference of two asterisks to the description of a new genus, so that this name together with Forsskål's generic description made a *descriptio generico-specifica,* in agreement with ICN, Art. 38.3. The name *Pteranthus dichotomus* Forssk. is therefore legitimate, and is the correct name for the species in *Pteranthus,* while Mirbel's name is an illegitimate renaming.
Correct orthography: *Pteranthus forskaolii* Mirb.
Nomenclatural status: Illegitimate.
Currently used name for the taxon: ***Pteranthus dichotomus*** Forssk.

Silene Forskohlei Steud., *Nomencl. Bot.*, ed. 2, 2: 584, 1841.
According to ICN, the orthography of the eponym must be corrected to *Silene forskohlii* Steud. – When publishing this eponym, Steudel cited a name ascribed to Forsskål, '*Silene diffusa* Forssk.', apparently in synonymy just below *S. forskohlii* Steud., and because the work was a nomenclator, there was no description or further synonymy. The name '*Silene diffusa* Forssk.' is not found in Forsskål's works, and not being a replacement name and without a description, *Silene forskohlii* Steud. is not validly published. *Silene diffusa* Otth, in DC., *Prodr.* 1: 373, 1824, is a later legitimate name of a plant described from Italy. One of our reviewers has suggested that this situation may have arisen as the result of Steudel's misreading of *Jussiaea diffusa* Forssk., *Fl. Aegypt.-Arab.*: LXVI, No. 235; 210, Cent. VIII, No. 45, 1775. On p. 210, *Jussiaea* (two species) and *Silene* (two species) occur together in the group *Decandria*, with *Jussiaea diffusa* just above the species of *Silene*, and Steudel may in his manuscript have wrongly listed *Jussiaea diffusa* as a species of *Silene*. However, *Jussiaea diffusa* Forssk. is also mentioned in Steudel's work, where it occurs on p. 836.
Correct orthography: *Silene forskohlii* Steud.
Nomenclatural status: Not validly published.
Currently used name for the taxon: No currently accepted name known; probably a lapsus by Steudel.

Celastraceae

Catha forskalii A.Rich., *Tent. Fl. Abyss.* 1: 134, 1847 [22 May 1847].
According to ICN, the eponym needs no correction. – In the protologue of his eponym, Richard indicated that this species is identical with Forsskål's *Catha* No. 4. This must refer to the circumstance that the description of the species *Catha* [without epithet] is number four in *Fl. Aegypt.-Arab.*: 63, *Cent.* III, No. 4, 1775. Forsskål's description and the genus name *Catha* are based on a plant from Yemen. At 'No. 4' on p. 63 in Forsskål's work, there is a generic name, *Catha*, and a description of a species without epithet. From the typography and lay-out it appears that the rank of the unnamed taxon is that of a species, not that of a genus, because immediately after it follows a similar description of a second species from Yemen that is named *C. spinosa* Forssk., *Fl. Aegypt.-Arab.*: CVII, No. 156; 64, *Cent.* III, No. 5, 1775. However, for 'No. 4' without epithet on p. 63, the binomial appears elsewhere in the *Fl. Aegypt.-Arab.* as *C. edulis* Forssk., *Fl. Aegypt.-Arab.*: CVII, No. 155, 1775. However, since Forsskål's work contains no separate generic diagnosis or description of *Catha* (required for the valid publication of a new genus by ICN, Art. 38.1), all species names under it are not validly published, unless the conditions for a *descriptio-generico specifica* (see Art. 38.5) are met, and this is not the case for a new genus with two new species. Scopoli validated the generic name *Catha* Forssk. ex Scop., *Intr. Hist. Nat.*: 228, 1777 [Jan.–Apr. 1777], and Vahl transferred Forsskål's two not validly published species in *Catha* to the genus *Celastrus* as *C. edulis* Vahl, *Symb. Bot.* 1: 21, 1790, and *C. parviflorus* Vahl, *Symb. Bot.* 1: 21, 1790. Richard's *Catha forskalii* is a validly published species name in a legitimate genus, but it is a superfluous, illegitimate renaming because it cites in synonymy the earlier legitimate name *Celastrus edulis* Vahl. The currently accepted names for Forsskål's two species in *Catha* are *C. edulis* (Vahl) Forssk. ex Endl., *Enchir. Bot.*: 575, 1841, and *Gymnosporia parvifolia* (Vahl) Chiov., in *Bull. Soc. Bot. Ital.* 1923(1–2): 114, 1923.
Correct orthography: *Catha forskalii* A.Rich.
Nomenclatural status: Illegitimate.
Currently used name for the taxon: ***Catha edulis*** (Vahl) Forssk. ex Endl.

Maytenus forsskaoliana Sebsebe, *Symb. Bot. Upsal.* 25(2): 67, 1985.
According to ICN, the eponym needs no correction. A new combination has recently been made, *Gymnosporia forsskaoliana* (Sebsebe) Jordaan, in *Taxon* 55(2): 521–522, 2006 [22 Jun. 2006]. – Sebsebe's eponym was given to a new taxon from Yemen; it is honorific without direct association with Forsskål's work or collections, except for its origin from Yemen. It is a validly published and legitimate name.

Correct orthography: *Maytenus forsskaoliana* Sebsebe
Nomenclatural status: Legitimate.
Currently used name for the taxon: ***Gymnosporia forsskaoliana*** (Sebsebe) Jordaan

Commelinaceae

Aneilema Forskâlei Kunth, *Enum. Pl.* 4: 71, 1843 [17–19 Jul. 1843]. The letter 'â' in the original publication must be a typographic error for 'å', the Scandinavian 'å' not being available in the typesetter's case.
According to ICN, the orthography of the eponym must be corrected to *Aneilema forskaolii* Kunth. – The species was first named by Forsskål as *Commelina tuberosa* Forssk., *Fl. Aegypt.-Arab.*: CIII, No. 39; 12, *Cent.* I, 33, 1775, based on a plant from Yemen, but there was already a *C. tuberosa* L., *Sp. Pl.* 1: 41, 1753, described from Central America. In agreement with other similar cases in Forsskål's work, we conclude that Forsskål's name is a misidentification, *C. tuberosa sensu* Forssk., non L. Vahl changed the name to *C. paniculata* Vahl, *Enum. Pl.* 2: 179, 1805, but that is an illegitimate later homonym of *C. paniculata* Hill, *Veg. Syst.* 23: 4, 1773. Therefore Kunth's eponym is the first legitimate name for Forsskål's plant.
Correct orthography: *Aneilema forskaolii* Kunth
Nomenclatural status: Legitimate.
Currently used name for the taxon: ***Aneilema forskaolii*** Kunth

Commelina Forskâlaei Vahl, *Enum. Pl.* 2: 172 [on p. 404 as Forskâlei], 1805. The letter 'â' in the original publication must be a typographic error for 'å,' which is how the eponym is spelt elsewhere in Vahl's work, and the epithet *Forskâlaei* is corrected to *Forskâlei* in the index at the end of Vahl's *Enum. Pl.* 2: 404 (1805).
And
Commelina Forskailii Hochst. ex C.B.Clarke, in A.DC. & C.DC., *Monogr. Phan.* 3: 173, 1881 [Jun. 1881].
According to ICN, the orthography of Vahl's eponym must be corrected to *Commelina forskaolii* Vahl. – Vahl described his species as new, based on material collected by Forsskål, citing "Habitat in Arabia. Forskål", but not citing any species name or description by Forsskål. Vahl's name is therefore valid and legitimate. Based on Vahl's annotations on the sheet, Hepper & Friis (1994: 255) have suggested that *Forsskål 35* (C barcode C10015141) is part of the original material and a suitable lectotype for *C. forskaolii* Vahl, but although this is the currently accepted name for this species, no formal lectotypification seems to have been made. There are a number of orthographic variants of *C. forskaolii* Vahl that it may be useful to discuss here. *Commelina forskaalii* Hochst. is a printed label-name for collection 105 in the *exsiccata*: "Kotschyi iter Nubicum / 105. *Commelina Forskaalii* Vahl. / Ad montem Cordofanum Arush Cool in lipidosis et arenosis. /U.i. 1841. D. 2. Oct. 1839." [In rocky and sandy places at the mountain Arush Cool in the Kordofan province. Distributed by the *Unio Itineraria* 1841 (see below). Collected on 2nd October 1839.] This specimen has been seen from Herb. HEID at *Global Plants*, https://plants.jstor.org/stable/10.5555/al.ap.specimen.heid701054. The specimens in the *iter Nubicum exsiccata* were collected by Theodor Kotschy in the former Kordofan Province, Central Sudan, and distributed in 1841 by C.F.F. Hochstetter (1787–1860) via *Unio Itineraria* (U.i. on label; in German, *Botanische Reiseverein*; Wörz 2007). The label is without descriptive text, and cites Vahl, so the plant name is an orthographic variant without nomenclatural standing. Later, in a work on plants collected by Ignatus Knoblecher in Sudan, Kotschy, in *Sitzungsber. Kaiserl. Akad. Wiss. Wien., Math.-Naturwiss. Cl.* 50(1–5): 353, 1865, listed the same collection under the name *Commelina Forskailii* Vahl. Finally, Clarke, in A.DC. & C.DC., *Monogr. phan.* 3: 173, 1881, published Kotschy's orthographic variant as "*Commelina Forskailii*, Hochst in herb. Kotschy, non *C. Forskalæi* Vahl". The orthographic variant was published in synonymy of *C. kotschyi* Hassk. (1867), now a taxonomic synonym of *C. imberbis* Ehrenb. ex Hassk., in Schweinf., *Beitr. Fl. Aethiop.*: 209, 1867, also with the exclusion of *C. forskaolii* Vahl (as *Forskålii*). This is one of the most notable examples of misidentification and misspelling of Forsskål eponyms.
Correct orthography: *Commelina forskaolii* Vahl
Nomenclatural status: Legitimate.

Currently used name for the taxon*: *Commelina forskaolii Vahl

Convolvulaceae

Convolvulus Forskalii Delile, *Descr. Egypte, Hist. Nat.*: 190, tab. 18, fig. 3, 1813 [text published 1813–1814; according to TL-2 the plates were only published in 1826].
According to ICN, the eponym should be written *Convolvulus forskalii* Delile. – In the protologue of his eponym, Delile cited as synonyms *C. cneorum* Forssk., *Fl. Aegypt.-Arab.*: LXIII, No. 124; CVI, No. 120, 1775, based on a plant from Egypt, and *C. lanatus* Vahl, *Symb. Bot.* 1: 16, 1790, the latter being a new name for *C. cneorum* Forssk., because there was already a species named *C. cneorum* L., *Sp. Pl.* 1: 157, 1753, described from the Central Mediterranean. We therefore consider Forsskål's *C. cneorum* a misidentification, *C. cneorum sensu* Forssk., non L. Vahl proposed the legitimate name *C. lanatus* Vahl for Forsskål's *Convolvulus cneorum*, for which reason Delile's name is a superfluous, illegitimate renaming.
Correct orthography: *Convolvulus forskalii* Delile
Nomenclatural status: Illegitimate.
Currently used name for the taxon*: *Convolvulus lanatus Vahl

Convolvulus Forskolei Spreng., *Syst. Veg.*, ed. 16, 1: 596. 1824 [dated 1825; publ. in late 1824].
And
Ipomoea Forskolei Steud., *Nomencl. Bot.*, ed. 2. 1: 816, 1840.
According to ICN, the orthography of these eponyms must be corrected to *Convolvulus forskolii* Spreng. and *Ipomoea forskolii* Steud. – In the protologue of his eponym, Sprengel cited in synonymy *Ipomoea verticillata* Forssk., *Fl. Aegypt.-Arab.*: CVI, No. 126; 44, *Cent.* II, No. 41, 1775, based on a plant from Yemen. Forsskål's name was cited unaltered by Vahl, in *Symb. Bot.* 3: 33, 1794, but Linnaeus had previously described a species named *I. verticillata* L., *Syst. Nat.*, ed. 10, 2: 924, 1759, on material from America. As in parallel cases, we conclude that Forsskål's name represents a misidentification, *I. verticillata sensu* Forssk., non L., and Sprengel therefore proposed a new name. However, there was also already a *Convolvulus forskalii* Delile (see previous name), which we have concluded is illegitimate because it is a superfluous renaming of *C. lanatus* Vahl, and one could therefore assume that *C. forskolii* Spreng. is legitimate. However, we are of the opinion that *C. forskalii* Delile (1813) and *C. forskolii* Spreng. (1824) are so similar that they are likely to be confused; they are therefore to be treated as homonyms in agreement with Art. 53.2 of the ICN, even when the earlier homonym was illegitimate (Art. 53.1. Note 2). Sprengel's name is therefore an illegitimate later homonym of Delile's (ICN Art. 53.2). Later again, Steudel published *Ipomoea forskolii* Steud., *Nomencl. Bot.*, ed. 2. 1: 816, 1840, citing *Ipomoea verticillata sensu* Forssk., non L., and *Convolvulus forskolii* A.Dietr. in synonymy. We have not traced the designation *Convolvulus forskolii* A. Dietr. as an independently published name in IPNI (continuously updated) or in the publications of Albert Gottfried Dietrich (A.Dietr.; 1795–1856). Because *Convolvulus forskolii* Spreng. is illegitimate, Steudel's name in a genus distinct from *Convolvulus*, is a legitimate new name for *Ipomoea verticillata sensu* Forssk., non L. Both *Convolvulus forskolii* Spreng. and *Ipomoea forskolii* Steud. are now considered taxonomic synonyms of *I. biflora* (L.) Pers., *Syn. Pl.* 1: 183, 1805, based on *C. biflorus* L., *Sp. Pl.*, ed. 2, 2: 1668, 1763. Forsskål listed *C. biflorus* in *Fl. Aegypt.-Arab.*: LXII, No. 122; 203, *Cent.* VIII, No. 9, 1775, from Egypt, on p. LXII the species name is provided with one asterisk, suggesting that it is intended to be a new species. Neither *C. biflorus* L., nor *I. biflora* (L.) Pers. are recorded from Egypt by Boulos (2000).
For ***Convolvulus Forskolei*** Spreng.
Correct orthography: *Convolvulus forskolii* Spreng.
Nomenclatural status: Illegitimate.
And for ***Ipomoea Forskolei*** Steud.
Correct orthography: *Ipomoea forskolii* Steud.
Nomenclatural status: Legitimate.
Currently used name for the taxon*: *Ipomoea biflora (L.) Pers.

Pharbitis Forskoeli G.Don, *Gen. Hist.* 4(1): 263, 1837. The letters 'oe' are in the original publication printed as a ligature, 'œ.'

According to ICN, the orthography of the eponym must be corrected to *Pharbitis forskoelii* G.Don. – In the protologue of his eponym, species of *Pharbitis* no. 9, Don cited *Ipomoea scabra* Forssk., *Fl. Aegypt.-Arab.*: CVI, No. 122; 44, *Cent.* II, No. 43, 1775, a legitimate name ("Native of Arabia, at *Hadie, Ipomoea scabra*, Forsk. aegypt. p. 44."). In an earlier publication, Gmelin, *Syst. Nat.*, ed. 13bis, 2(1): 345, 1791, had cited and republished Forsskål's name, *I. scabra*, unchanged and entirely based on a reference to Forsskål. Don, on the other hand, apart from publishing the new name, *P. forskoelii* G.Don, on the same page also transferred Gmelin's *I. scabra* to *Pharbitis* as *P. scabra*, species no. 13, referring to the description of the name *Ipomoea scabra* Gmel. as given in and by Cavanilles, *Descr. Pl.*: 101, 1801, and indirectly citing Forsskål's *I. scabra*. Don has therefore proposed both a new name, the Forsskål eponym, and a new combination using Forsskål's original epithet for the two species on the same page, that in *Gen. Hist.* 4: 263, 1837. We do not find that the ICN guides us in this situation, as both names are ultimately based on Forsskål's *I. scabra*. One of our reviewers argues that both Don's names are legitimate because *Pharbitis scabra* is not based on *Ipomoea scabra* Forssk. for the reason that G.Don used that name as the basis for his *Pharbitis forskoelii,* while his *Pharbitis scabra* is based on references to Gmelin and Cavanilles. We are of the opinion that the situation is caused by a simple oversight by Don. The two names are now considered synonyms, whether legitimate or illegitimate, of the variable and widely cultivated *Ipomoea nil* (L.) Roth, *Catal. Bot.* 1: 36, 1797.
Correct orthography: *Pharbitis forskoelii* G.Don
Nomenclatural status: Legitimate?
Currently used name for the taxon: ***Ipomoea nil*** (L.) Roth

Cucurbitaceae

Luffa Forskalii Beck & F.Abel in *Wiener Ill. Gart.-Zeitung* 16: 451, illustration on p. 452, 1891 [Nov. 1891].
And
Luffa Forskalii hort. ex *Bull. Soc. Tosc. Ortic.* 1893: 18: 12, fig. 1, 1893 [abbreviation of title of journal according to BPH (continuously updated)].
And
Luffa Forskalii Schweinf. ex Harms in *Repert. Spec. Nov. Regni Veg.* 19: 232, 1923.
According to ICN, these eponyms should be written *Luffa forskalii* Beck & F.Abel. The three publications do not refer to each other, the oldest publication is based on plants grown in Vienna, the second on plants grown in Tuscany, and the third based on collections from nature. We therefore consider the name from Tuscany either an illegitimate homonym of *L. forskalii* Beck & F.Abel or simply a name cited with no intention of a nomenclatural act. Harms' name is for a new species, based on *Schweinfurth 545* (duplicate in the Grey Herbarium; G00031905), from the Tihama lowlands of Yemen, and with the label name "*Luffa Forskali* Schweinfurth mscr." Harms states that "Die neue Art, die Schweinfurth zu Ehren von P. Forskål benannt hat, ist nahe verwandt mit *Luffa cylindrica* (L.) Roem. und *L. acutangula* (L.) Roxb. ..." [The new species, which Schweinfurth has named in the honour of P. Forskål [sic!], is closely related to *Luffa cylindrica* (L.) Roem. and *L. acutangula* (L.)Roxb.] – In the protologue of their eponym, Beck and Abel stated that they had acquired living material of their new species from *glücklichen Arabien* through the trading company Dammann & Co. There is still a company named Dammann that has been active in the import of tea and coffee since ca. 1700, presumably opening trade with coffee-producing Yemen after 1862, when the Suez Canal opened. The description of *L. forskalii* Beck & F.Abel is in German and the name is validly published, although in the style of popular horticultural journals; the same illustration is repeated in *Bull. Reale Soc. Tosc. Ortic.* 18, 1893, where it is stated that the species was first collected by Forsskål. From the

two published drawings, the plant can be identified as *L. acutangula* (L.) Roxb., *Fl. Ind.*, ed. 1832, 3: 713, 1832, based on *Cucumis acutangulus* L., *Sp. Pl.* 2: 1011, 1753., an Indian plant known to have been cultivated in Yemen, but Hepper & Friis (1994: 147–151) found no text in Forsskål's *Flora Aegyptiaco-Arabica* associated with this species and traced no Forsskål material that agrees with Beck and Abel's illustration and description. A specimen in the Forsskål Herbarium at C (numbered *Forsskål 1725*) has been identified by Charles Jeffrey as *Sicyos angulatus* L. (an American species). This specimen is probably not collected by Forsskål; Vahl noted on the sheet that he had received it from Attilo Zuccagni (1754–1807), director of the botanical garden and herbarium in Firenze.
Correct orthography: *Luffa forskalii* Beck & F.Abel, etc.
Nomenclatural status: Legitimate.
Currently used name for the taxon: ***Luffa acutangula*** (L.) Roxb.

Cyperaceae

Cyperus forskalianus Väre & Kukkonen, *Ann. Bot. Fenn.* 42(6): 477–478, fig. 4, 2005 [19 Dec. 2005].
According to ICN, the orthography of the eponym is correct. – The name was given to a new, recently discovered species from Yemen; it has no association with Forsskål except for its origin from the country where he collected, and for the fact that the authors were from Finland, where Forsskål was born in Helsinki.
Correct orthography: *Cyperus forskalianus* Väre & Kukkonen
Nomenclatural status: Legitimate.
Currently used name for the taxon: ***Cyperus forskalianus*** Väre & Kukkonen

Cyperus Forskålei A.Dietr., *Sp. Pl.*, ed. 6, 2: 251, 1833.
The letter interpreted here as an 'å' is in the original publication an 'o' with a smaller 'o' above. There is no letter exactly like this in any European alphabet we know, but it must be treated in the same way as an identical letter in *Delphinium grandiflorum* Forssk., where 'Forskol' is spelt with a small 'o' over the large second 'o,' where it obviously means 'å,' the Scandinavian 'å' not being available in the typesetter's case (Fig. 11).

According to the ICN, the orthography of the eponym must be corrected to *Cyperus forskaolii* A.Dietr. – In the protologue of his eponym, Dietrich cited in synonymy *C. ferrugineus* Forssk., *Fl. Aegypt.-Arab.*: LX, No. 20; 14, *Cent.* I, No. 43, 1775, based on a plant from Egypt, but there was an older, valid and legitimate name based on material from America, *C. ferrugineus* L., *Sp. Pl.* 1: 44, 1753. We conclude that *C. ferrugineus* Forssk. is a misidentification, *C. ferrugineus sensu* Forssk., non L. For this reason, *C. forskaolii* A.Dietr. is a legitimate new name. It is currently considered a taxonomic synonym of *C. fuscus* L., *Sp. Pl.* 1: 46 (1753).
Correct orthography: *Cyperus forskaolii* A.Dietr.
Nomenclatural status: Legitimate.
Currently used name for the taxon: ***Cyperus fuscus*** L.

Ericaceae

Erica Forskalii Vitman, *Summa Pl.* 2: 414, 1789–1790 [Dec. 1789 or early 1790].
According to ICN, the eponym should be written *Erica forskalii* Vitman. – In the protologue of his eponym, Vitman cited in synonymy *E. verticillata* Forssk., *Fl. Aegypt.-Arab.*: XXV, No. 187; 210, *Cent.* VIII, No. 43, 1775, based on a plant collected near Constantinople (Istanbul). However, Bergius, not knowing the contents of Forsskål's manuscripts, had shortly after Forsskål's death in Yemen published *E. verticillata* P.J.Bergius, *Descr. Pl. Cap.*: 99, 1767, for a plant from South Africa. Forsskål's name is therefore an illegitimate later homonym, and Vitman's name is a legitimate replacement name. Salisbury, in the protologue of *E. manipuliflora* Salisb., in *Trans. Linn. Soc. London* 6: 344, 1802, cited in synonymy Forsskål's illegitimate name *E. verticillata* Forssk. ('p. 210'), but Vitman's name has priority over Salisbury's, and consequently *E. forskalii* Vitman should have been taken up. Thus, *E. manipuliflora* Salisb. is a superfluous, illegitimate renaming of *E. forskalii* Vitman. However, *E. manipuliflora* Salisb. was conserved by the International XVIII Botanical Congress in Melbourne against *E. Forskalii*

Vitman, according to a proposal by Nelson & Wulff, in *Taxon* 56: 959, 2007.
Correct orthography: *Erica forskalii* Vitman
Nomenclatural status: Legitimate when published, rejected in favour of *Erica manipuliflora* Salisb.
Currently used name for the taxon: ***Erica manipuliflora*** Salisb.

Euphorbiaceae

Adenoropium Forskålii Pohl, *Pl. Bras. Icon. Descr.* 1: 15, 1826.
According to ICN, the orthography of the eponym must be corrected to *Adenoropium forskaolii* Pohl. (Fig. 9). – In the protologue, Pohl proposed *Adenoropium Forskålii* Pohl in what was then a new genus, and he cited two synonyms of his new species, *Jatropha glandulosa* Vahl and *Croton villosus* Forssk. [as *villosum*]. The earliest of these, *Croton villosus* Forssk., *Fl. Aegypt.-Arab.*: CXXII, No. 563; 163, *Cent.* VI, No. 27, 1775, [as *villosum*], based on a plant from Yemen, was legitimate but not placed in the correct genus. Vahl transferred Forsskål's species to the Linnaean genus *Jatropha* as *J. glandulosa* Vahl, *Symb. Bot.* 1: 80, 1790, thereby creating a superfluous name. Pohl's *Adenoropium forskaolii* was not illegitimate because, on the same page, he described the new species *A. villosum* Pohl, from Brazil, thereby precluding the transfer of Forsskål's epithet to *Adenoropium*. Realizing the priority of Forsskål's *Croton villosus,* accepting its position in the genus *Jatropha*, and citing Pohl's name *A. forskaolii* in synonymy, Müller Argoviensis transferred Forsskål's epithet to *Jatropha* as *J. villosa* (Forssk.) Müll.Arg., in A.DC., *Prodr.* 15(2): 1085, 1866. However, he did this without taking note of the by then already validly and legitimately published *J. villosa* Wight, in Wight, *Icon. Pl. Ind. Orient.* 4(1): 1, 1848 [text to Wight's plate 1169; a replacement name for *J. peltata* Wight, the name which appears on plate 1169, not *J. peltata* Kunth in Humb. *et al.*, *Nov. Gen. Sp.* 2(pt. 6): 83 (ed. fol.); 104 (ed. qto.), 1817, nor *J. peltata* Sessé in Cerv., *Supl. Gaz. Lit. Mexico* 3 (suppl.): 3, adnot. no. 6, 1794, and thereby creating an illegitimate later homonym. Finally, Courbon described a species from the Dahlak Archipelago in the Red Sea, *J. pelargoniifolia* Courbon, in *Ann. Sci. Nat., Bot.*, Sér. 4, 18: 150, 1862, a species that according to present taxonomy belongs to the same species as *A. forskaolii* Pohl, *J. glandulosa* Vahl and *C. villosus* Forssk. *Adenoropium forskaolii* Pohl (1826) is legitimate according to ICN, Art. 11.4. Note 2, Ex. 18, and has priority over the widely accepted *Jatropha pelargoniifolia* Courbon (1862), although it has not been used for nearly 200 years. *Adenoropium forskaolii* Pohl has been proposed for rejection, with *Jatropha pelargoniifolia* Courbon conserved against it (Friis & Thulin 2023).
Correct orthography: *Adenoropium forskaolii* Pohl
Nomenclatural status: Legitimate.
Currently used name for the taxon: ***Jatropha pelargoniifolia*** Courbon with *Adenoropium forskaolii* Pohl proposed for rejection.

Euphorbia Forskålii J.Gay, in Webb & Berth., *Hist. Nat. Iles Canaries (Phytogr.)* 3 (pt. 2, sect. 3): 240, 1847 [Apr.–May 1837].
And
Euphorbia Forskalei Bourg. ex Boiss., in A.DC., *Prodr.* 15(2.1): 47, 1862 [late Jan. 1862].
According to ICN, the orthography of Gay's eponym must be corrected to *Euphorbia forskaolii* J.Gay and the second by Boissier to *E. forskalii* Bourg. ex Boiss. (Fig. 9). – In the protologue of his eponym, Gay cited in synonymy *E. thymifolia* Forssk., *Fl. Aegypt.-Arab.*: CXII, No. 310; 94, *Cent.* VI, No. 93, 1775, based on a plant of unspecified origin, but collected on the Royal Danish Expedition to Egypt and Yemen. However, there was already a validly published and legitimate *E. thymifolia* L., *Sp. Pl.* 1: 454, 1753, described from America. We conclude that Forsskål's name is a misidentification, *E. thymifolia sensu* Forssk., non L. Therefore, *E. forskaolii* J.Gay is a legitimate new name. *Euphorbia forskaolii* J.Gay is currently accepted, but various new combinations in genera split-off from *Euphorbia* are based on Gay's name without always maintaining the transcription of 'å' to 'ao' proscribed in the ICN: *Anisophyllum forskaolii* [as *forskalei*] (J.Gay) Klotzsch & Garcke in *Abh.*

Königl. Akad. Wiss. Berlin 1859: 25, 1860, *Chamaesyce forsskaolii* (J.Gay) Figueiredo, *Fl. Cabo Verde* 54: 48, 1996, and *C. forskaolii* [as *forskalii*] (J.Gay) Parolly in *Willdenowia* 37(1): 251, 2007 [31 Aug. 2007]. *Euphorbia forskaolii* J.Gay is similar to *E. scordifolia* Jacq., *Icon. Pl. Rar.* 3: 5, 1793, and the two species have partly overlapping distributions; *E. scordifolia* Jacq. has priority if the two taxa are considered conspecific. Boissier's eponym, *Euphorbia forskalii* Bourg. ex Boiss. is published in synonymy of the accepted name for that species, *E. prostrata* Aiton, *Hort. Kew.* 2: 139, 1789, and is not validly published.

For ***Euphorbia Forskålii*** J.Gay
Correct orthography: *Euphorbia forskaolii* J.Gay
Nomenclatural status: Legitimate.
Currently used name for the taxon: ***Euphorbia forskaolii*** J.Gay

For ***Euphorbia Forskalei*** Bourg. ex Boiss.
Correct orthography: *Euphorbia forskalii* Bourg. ex Boiss.
Nomenclatural status: Not validly published.
Currently used name for the taxon: ***Euphorbia prostrata*** Aiton

Phyllanthus forskahlii Lepr. ex Baill., *Recueil Observ. Bot.* 1: 59, 1860.
According to ICN, the orthography of the eponym is correct. – Baillon cited Leprieur's label name in synonymy of *Euphorbia forskaolii* [as *forskålii*] J.Gay; the name is therefore not validly published.
Correct orthography: *Phyllanthus forskahlii* Lepr. ex Baill.
Nomenclatural status: Not validly published.
Currently used name for the taxon: ***Euphorbia forskaolii*** J.Gay

Fabaceae (Leguminosae in Hepper & Friis 1994)

Astragalus Forskahlei Boiss., *Diagn. Pl. Orient.*, Ser. 1, 9: 101, 1849 [Jan.–Feb. 1849].
According to ICN, the orthography of the eponym must be corrected to *Astragalus forskahlii* Boiss. Kuntze proposed a new combination based on this name, *Tragacantha forskahlii* [as *Forskalei*] (Boiss.) Kuntze, *Revis. Gen. Pl.* 2: 945, 1891 [5 Nov. 1891]. – In the protologue of his eponym, Boissier cited in synonymy *Colutea spinosa* Forssk., *Fl. Aegypt.-Arab.*: LXX, No. 359; 131, *Cent.* V, No. 7, 1775, based on a plant from Egypt. Forsskål's name is validly published and legitimate, and it has been transferred to *Astragalus* as *A. spinosus* (Forssk.) Muschl., in *Verh. Bot. Vereins Prov. Brandenburg* 49: 98, 1907. Christensen overlooked Muschler's transfer and published the isonym *A. spinosus* (Forssk.) C.Chr., in *Dansk Bot. Ark.* 4(3): 24, 1922. *Astragalus forskahlii* Boiss. and *T. forskahlii* (Boiss.) Kuntze are superfluous, illegitimate changes of epithet.
Correct orthography: *Astragalus forskahlii* Boiss.
Nomenclatural status: Illegitimate.
Currently used name for the taxon: ***Astragalus spinosus*** (Forssk.) Muschl.

Cassia Forskalii Royle, *Man. Mat. Med.*: 356, 1847.
According to ICN, the eponym should be written as *Cassia forskalii* Royle. – In the protologue of his eponym, Royle cited in synonymy *C. lanceolata* Forssk., *Fl. Aegypt.-Arab.*: CXI, No. 270; 85, *Cent.* III, No. 85, 1775, based on a plant from Yemen. Forsskål's name is legitimate and later placed as a taxonomic synonym of *Senna sophera* (L.) Roxb., *Fl. Ind.*, ed. 1832, 2: 347, 1832, based on *Cassia sophera* L, *Sp. Pl.*: 379, 1753. Royle's name therefore represents a superfluous, illegitimate change of epithet.
Correct orthography: *Cassia forskalii* Royle
Nomenclatural status: Illegitimate.
Currently used name for the taxon: ***Senna sophera*** (L.) Roxb.

Lupinus Forskahlei Boiss., *Diagn. Pl. Orient.*, Ser. 1, 9: 10, 1849 [Jan.–Feb. 1849].
According to ICN, the orthography of the eponym must be corrected to *Lupinus forskahlii* Boiss. – In the protologue of his eponym, Boissier cited in synonymy *L. digitatus* Forssk., *Fl. Aegypt.-Arab.*: LXX, No. 358; 131, *Cent.* 5, No. 6, 1775, based on a plant from Egypt. Boissier published his name in a note below a diagnosis of the unrelated new species *L. palaestinus* Boiss. However, Forsskål's name is validly published

and legitimate, and Boissier's name represents a superfluous, illegitimate change of epithet.
Correct orthography: *Lupinus forskahlii* Boiss.
Nomenclatural status: Illegitimate.
Currently used name for the taxon: ***Lupinus digitatus*** Forssk.

Lamiaceae (Labiatae in Hepper & Friis 1994)

Ballota Forskahlei Benth., *Labiat. Gen. Spec.*: 599, 1834 [Aug. 1834].
According to ICN, the orthography of the eponym must be corrected to *Ballota forskahlii* Benth. – In the protologue of his eponym, Bentham cited in synonymy *Phlomis alba* Forssk., *Fl. Aegypt.-Arab.*: CXIV, No. 355; 107, *Cent.* IV, No. 26, 1775, based on a plant from Yemen. Forsskål's name is legitimate, but the earlier legitimate name *B. alba* L., *Fl. Suec.*, ed. 2: 206, 1755, prevents transfer of Forsskål's epithet to the Linnaean genus *Ballota*. Bentham's new epithet is therefore a legitimate renaming. Sebald transferred Forsskål's species to *Leucas* as *L. alba* (Forssk.) Sebald, in *Stuttgarter Beitr. Naturk.*, A 308: 38, 1978, which is legitimate.
Correct orthography: *Ballota forskahlii* Benth.
Nomenclatural status: Legitimate.
Currently used name for the taxon: ***Leucas alba*** (Forssk.) Sebald

Micromeria Forskahlei Benth., *Labiat. Gen. Spec.*: 379, 1834 [May 1834].
According to ICN, the orthography of the eponym must be corrected to *Micromeria forskahlii* Benth. – In the protologue of his eponym, Bentham cited in synonymy *Thymus imbricatus* Forssk., *Fl. Aegypt.-Arab.*: CXIV, No. 360; 108, *Cent.* IV, No. 29, 1775, based on a plant from Yemen. Forsskål's name is legitimate, and Bentham's name is therefore a superfluous, illegitimate renaming. Vahl, in *Symb. Bot.* 2: 65. 1791, identified Forsskål's species as *T. piperella* L., *Syst. Nat.*, ed. 12, 2: 400, 1767, but Briquet did not accept that identification and transferred Forsskål's name to the Linnaean genus *Satureja* as *S. imbricata* (Forssk.) Briq., in Engl. & Prantl, *Nat. Pflanzenfam.* IV(3a): 299, 1896. Christensen later transferred Forsskål's name to *Micromeria* as *M. imbricata* (Forssk.) C.Chr., in *Dansk Bot. Ark.* 4(3): 21, 1922.
Correct orthography: *Micromeria forskahlii* Benth.
Nomenclatural status: Illegitimate.
Currently used name for the taxon: ***Micromeria imbricata*** (Forssk.) C.Chr. or ***Satureja imbricata*** (Forssk.) Briq.

Moscharia Forskåhlii J.F.Gmel., *Syst. Nat.*, ed. 13bis, 2(1): 275, 1791 [late Sep.–Nov. 1791]. The sign over the 'a' in the original publication is part of an 'o' and the letter must be interpreted as an 'å,' the Scandinavian 'å' not being available in the typesetter's case (Fig. 10). According to ICN, the orthography of the eponym must be corrected to *Moscharia forskaohlii* J.F.Gmel. – In the protologue of his eponym, Gmelin cited Forsskål's *Fl. Aegypt.-Arab.*: 158, *Cent.* VI, No. 10, 1775, which is a reference to a description of a species of *Moscharia* without a specific epithet, based on a plant from Egypt. However, Gmelin did not note that Forsskål's species had been named in the *Florulae* as *M. asperifolia* Forssk. on p. LXXIV, No. 472, which means that Gmelin's name represents a superfluous, illegitimate change of epithet. Vahl, in *Symb. Bot.* 1: 40, 1790, identified *M. asperifolia* Forssk. with *Teucrium iva* L., *Sp. Pl.* 2: 563, 1753. *Teucrium iva* L. is now transferred to the genus *Ajuga* as *A. iva* (L.) Schreb., *Pl. Verticill. Unilab. Gen. Sp.*: xxv (1773).
Correct orthography: *Moscharia forskaohlii* J.F.Gmel.
Nomenclatural status: Illegitimate.
Currently used name for the taxon: ***Ajuga iva*** (L.) Schreb.

Ocimum Forskölei Benth., *Labiat. Gen. Spec.*: 6, 1832 [Aug. 1832]. The use of the letter 'ö' must be a typographical error for 'å,' the Scandinavian 'å' not being available in the typesetter's case for 'å'.
According to ICN, the orthography of the eponym must be corrected, with 'ö'/'å' transcribed to 'ao'. Wood, in *Kew Bull.* 37(4): 601, 1983, corrected *O. Forskölei* Benth. to *Ocimum forskolei* Benth., but with the typographic error corrected, it has to become *O. for-*

skaolii Benth. – In the protologue of his eponym, Bentham cited in synonymy *O. gratissimum* Forssk., *Fl. Aegypt.-Arab.*: CXIV, No. 363; 110, *Cent.* IV, No. 35, 1775, based on a plant from Yemen, but pointed out that this was not identical with the earlier published *O. gratissimum* L., *Sp. Pl.* 2: 1197, 1753, a tropical Old World species stated by Linnaeus to occur in India. We conclude that Forsskål's name is a misidentification, *Ocimum gratissimum sensu* Forssk., non L., and Bentham's new name is therefore validly published and legitimate.
Correct orthography: *Ocimum forskaolii* Benth.
Nomenclatural status: Legitimate.
Currently used name for the taxon: ***Ocimum forskaolii*** Benth.

Plectranthus Forskålæi Vahl, *Symb. Bot.* 1: 44, 1790. According to ICN, the orthography of the eponym must be corrected to *Plectranthus forskaolii* Vahl. Wood, in *Kew Bull.* 37(4): 599, 1983, transcribed this name as *P. Forsskalaei* Vahl, changing the eponym to a spelling with double 's', but not transcribing 'å' to 'ao' and not correcting 'æi' to 'ii.' – In the protologue of his eponym, Vahl cited in synonymy *Ocimum hadiense* Forssk., *Fl. Aegypt.-Arab.*: CXV, No. 367 [here reproduced in Fig. 5]; 109, *Cent.* IV, No. 32, 1775, based on a plant from Yemen. However, *O. hadiense* Forssk. is a validly published and legitimate name, and Vahl's name is therefore a superfluous, illegitimate change of epithet. Willdenow adopted the spelling '*Plectranthus Forskolæi*' in *Sp. Pl.*, ed. 4, 3(1): 169, 1800, thereby producing an orthographic variant of Vahl's name. Poiret transferred Vahl's illegitimate epithet to Lamarck's genus *Germanea* as *G. forskaolii* [as *Forskolæi*] Poir., *Encycl.*, Suppl. 2: 764, 1812, which is also illegitimate and attributed to Poiret alone. Kuntze transferred Willdenow's orthographic variant of the epithet to the genus *Majana* Rumph., *Herb. Amboin.* 5: 291, 296, t. 101, 13, 1747 [pre-Linnaean]. ex Bosc, in Déterv. (ed.), *Nouv. Dict. Hist. Nat.*, ed. 1, 13: 556, 1803, making the combination *Majana forskaolii* [as *Forskalii*] Kuntze, *Revis. Gen. Pl.* 2: 524, 1891 [5 Nov. 1891], again illegitimate and attributed to Kuntze alone. Briquet transferred this eponym to the genus *Coleus*, proposing the combination *C. forskaolii* [the combination applied the original orthographic variant of the eponym, *Forskålæi*] Briq., in Engler & Prantl, *Nat. Pflanzenfam.* 4(3A): 359, 1897, the latest illegitimate combination, attributed to Briquet alone. Schweinfurth used the correct *Plectranthus hadiensis* on labels of the specimens he distributed after his expedition to Yemen (see Christensen, 1918: 86), but the valid publication of the name was only provided by Sprenger in 1894 as *P. hadiensis* (Forssk.) Schweinf. ex Sprenger, in *Wiener Ill. Gart.-Zeitung* 19: 2, 1894, where he erroneously stated that the plant had come from Ethiopia. The species is now transferred to *Coleus*, as *C. hadiensis* (Forssk.) A.J.Paton, in *PhytoKeys* 129: 54, 2019.
Correct orthography: *Plectranthus forskaolii* Vahl
Nomenclatural status: Illegitimate.
Currently used name for the taxon: ***Coleus hadiensis*** (Forssk.) A.J.Paton

Salvia Forskaehlei L., *Mant. Pl.* [undated version, possibly a preprint].
And
Salvia Forskåhlei L., *Mant. Pl.*: 26, 1767 [15–31 Oct. 1767].
And
Salvia Forskohlei L., *Syst. Nat.*, ed. 12, 2: 67, 1767 [15–31 Oct. 1767].
According to ICN, the orthography of Linnaeus's different eponyms of this species must be corrected, respectively, to *Salvia forskaehlii* L., *S. forskaohlii* L., and *S. forskohlii* L. According to Stearn's introduction to the facsimile edition of the *Mantissa* (Stearn 1961), 'the text of the *Mantissa* ... interlocks with that of *Systema Naturae*, 12th ed., vol. 2, and Linnaeus's new genera and species were simultaneously published in both.' We therefore regard *S. forskaehlii* L., *S. forskaohlii* L. and *S. forskohlii* L. as simultaneously published orthographic variants. Two versions of the *Mantissa* exist, one of these without year of publication and without information on printers. This seems to be a kind of preprint or proofs, probably made to enable the citation of page number in *Systema Naturae*. Moreover, in this preprint, the existence of which

was not mentioned by Stearn (1961), Forsskål's name is misspelt as 'Forskähl', the species of *Salvia* as '*S. forskaehlei*', and Linnaeus's own name as 'Linée' on the title page. These, and other errors, were corrected in the final version of the *Mantissa*, which also has information on year of publication and printers on the title page. Unfortunately, the preprint is the version to which the names in IPNI (continuously updated) are linked at the time when we write this, which is why the spelling "*S. forskaehlei*" for the *Salvia* is given. That Linnaeus would have deliberately used the spellings 'Forskähl' and 'Linée' is unthinkable, and this preprint should not be used for nomenclatural purposes. In Linnaeus's own library, available on the internet through the Linnean Society, there are three copies of the *Mantissa*, BL.20/2A (https://linnean-online.org/119979/#?s=0&cv=4), BL.59A (https://linnean-online.org/120012/#?s=0&cv=0), and BL.60A (https://linnean-online.org/120014/#?s=0&cv=4&z=-0.1003%2C0%2C1.2005%2C1.4975), all of the published version. In *Syst. Nat.*, ed. 12, 2: 67, 1767, only a brief description of *S. Forskohlei* is given, along with a reference to 'Mant. [p.] 26', where a full description is found, along with information on collector ('Forskåhl') and area of origin ('Oriente'), and 'H.U.', indicating that the plant was grown in Hortus Upsaliensis. Gmelin, in the next edition of *Systema Naturae*, published after the death of Linnaeus, cited only the finally published *Mantissa* and the orthographic variant *S. Forskåhlii* (Gmelin, *Syst. Nat.*, ed. 13, 2(1) 47, 1791), with the lower half of a 'u' over the 'a'. We conclude that Gmelin's choice of orthographic variant should be followed (ICN, Art. 53.5, although this case is not exactly covered by the wording in the ICN). – The Linnaean eponyms of this species of *Salvia* were based on material grown in Uppsala from seeds sent by Forsskål from Constantinople (Istanbul). Forsskål also drafted a description of this plant in the field and named it *S. bifida* Forssk., *Fl. Aegypt.-Arab.*: XVIII, No. 13; 202, *Cent.* VIII, No. 2, 1775, a validly published and legitimate name, but the Linnaean name has priority. Rafinesque transferred the Linnaean species to *Terepis* [as *T. forskalei*] (L.) Raf., *Fl. Tellur.* 3: 94, 1837 [1836 publ. Nov.–Dec. 1837].

In cultivation, this species seems to be best known under the orthographic variant *Salvia forsskaolii* L., but use of the corrected Gmelin-variant ***Salvia forskaohlii*** L. seems to be in best agreement with the ICN.
Correct orthography: *Salvia forskaohlii* L.
Nomenclatural status: Legitimate.
Currently used name for the taxon: ***Salvia forskaohlii*** L.

Malpighiaceae

Cavcanthvs Forskahlii Raeusch., *Nomencl. Bot.*, ed. 3. 132, 1797.
According to ICN, the eponym should be written *Caucanthus forskahlii* Raeusch. – In his list of names in the *Nomenclator*, Räuschel only listed the name *Cavcanthvs Forskahlii* with the place of origin 'Arab[ia]' and the conventional sign for shrub habit. This is an indirect reference to *Caucanthus* Forssk., *Fl. Aegypt.-Arab.*: 91, *Cent.* III, No. 78, 1775, based on a plant from Yemen. On p. 91 in *Fl. Aegypt.-Arab.* there is a generic name and a description, but no specific epithet. However, the full species name, *C. edulis* Forssk., *Fl. Aegypt.-Arab.*: CXI, No. 289, 1775, is found elsewhere in the book with two asterisks, making up all the requirements for a *descriptio generico-specifica* in the ICN, Art. 38.5. In spite of the incomplete and indirect references, we consider Räuschel's name a superfluous, illegitimate change of epithet.
Correct orthography: *Caucanthus forskahlii* Raeusch.
Nomenclatural status: Illegitimate.
Currently used name for the taxon: ***Caucanthus edulis*** Forssk.

Moraceae

Cosaria Forskåhlii J.F.Gmel., *Syst. Nat.*, ed. 13bis, 2(1): 71, 1791 [late Sep.–Nov. 1791]. The sign over the 'a' in the original publication is part of an 'o' or an 'u' and the letter must be interpreted as an 'å,' the Scandinavian 'å' not being available in the typesetter's case (Fig. 10).
According to ICN, the orthography of the eponym must be corrected to *Kosaria forskaohlii* J.F.Gmel. –

Forsskål's name of the genus was *Kosaria* Forssk., *Fl. Aegypt.-Arab.*: 164, Cent. VI, No. 34, 1775, without a species name, and with the name of the species published elsewhere in the book, *Kosaria foetida* Forssk., *Fl. Aegypt.-Arab.*: CXXI, No. 532, 1775 [here reproduced in Fig. 6], making up all the requirements for a *descriptio generico-specifica* as defined in the ICN, Art. 38.5, and with two asterisks to mark the presence of a description of a new genus elsewhere in the book. Therefore, both the genus name and the species name are validly published and legitimate. Forsskål's name is based on a plant from Yemen. Gmelin published the genus name *Cosaria*, in *Syst. Nat.*, ed. 13bis, 2(1): 22, 1791 [late Sep.–Nov. 1791] with a diagnosis, but without reference to Forsskål: "COSARIA *Cal.[lyx] Communis* flores confertos in receptaculo plano includens." [COSARIA, with common calyx and the congested flowers included in a flat receptacle]. The text in that line concludes with the words *Carex diandra acuti*, part of a different plant name in the genus *Carex* and a term. This is probably the result of a typographical error, in a place where one would expect a reference to Forsskål. Gmelin's species name, *Cosaria forskåhlii*, is published separately in the same work, on p. 71 with reference to 'Forsk., Fl. Aeg. Arab., p. 164.' According to ICN, the eponym must be corrected to *Kosaria forskaohlii* J.F.Gmel., although Gmelin used an orthographic variant with a 'C' in the genus name instead of the original 'K'. Steudel produced another orthographic variant, *Korsaria* Forssk., by inserting a letter 'r', and also changed the orthography of Gmelin's epithet, in his eponym *Korsaria forskohlii* [as *Forskohlei*] (J.F.Gmel.) Steud., *Nomencl. Bot.*, ed. 2, 1: 850, 1840. – Gmelin cited in the protologue of his eponym a reference to Forsskål, *Fl. Aegypt.-Arab.*: 164 [here reproduced in Fig. 8], 1775; where the validly published and legitimate generic name, *Kosaria*, and a description can be found, but no specific epithet. Gmelin overlooked that Lamarck had already noted Forsskål's spelling of his genus name as *Kosaria* on p. 164 in *Fl. Aegypt.-Arab.*, and in Forsskål's *Ic. rer. nat.* (1776), where one can find the name and this comment by the editor (probably Zoëga) in the legend to the illustration on Tab. 20: "Ad genus *Dorstenia* L. referenda videtur." Lamarck published the name *Dorstenia radiata* Lam., *Encycl.* 2(1): 318, 1786, for what he thought was Forsskål's unnamed species of *Dorstenia* L., *Sp. Pl.*,121, 1753. Therefore, Lamarck's and Gmelin's names represent superfluous, illegitimate changes of Forsskål's valid epithet *foetida*. *Kosaria* is now the name of an infrageneric taxon in *Dorstenia, D.* sect. *Kosaria* (Forssk.) Engl. The name, *D. foetida* (Forssk.) Schweinf., *Bull. Herb. Boiss.* 4, Append. 2: 120 (1896), is the currently used for the species.
Correct orthography: *Kosaria forskaohlii* J.F.Gmel.
Nomenclatural status: Illegitimate.
Currently used name for the taxon: ***Dorstenia foetida*** (Forssk.) Schweinf.

Ficus Forskålaei Vahl, *Enum. Pl.* 2: 196, 1805.
According to ICN, the orthography of the eponym must be corrected to *Ficus forskaolii* Vahl. – In the protologue of his eponym, Vahl cited in synonymy *F. morifolia* Forssk., *Fl. Aegypt.-Arab.*: CXXIV, No. 622; 179, *Cent.* VI, No. 94, 1775, based on a plant from Yemen. Forsskål's name is validly published and legitimate and antedates the homonym *F. morifolia* Lam., *Encycl.* 2(2): 499, 1788, based on material collected from Réunion, and *F. morifolia* Vahl, Enum. Pl. 2: 203, 1805, based on material collected by J.G. König in India. *Ficus forskaolii* Vahl is therefore a superfluous, illegitimate change of epithet. The identity of *F. morifolia* has been doubtful, and Hepper & Friis (1994: 205) cited its identity as "*Ficus* sp., possibly young stage of *F. palmata* Forssk." and an annotation on the only extant material of *F. forskaolii* Vahl by Corner: "*Ficus palmata* v. young?" Currently, the name is generally accepted as being a synonym of *F. palmata* Forssk., *Fl. Aegypt.-Arab.*: CXXIV, No. 623; 179, *Cent.* VI, No. 95, 1775, also based on a plant from Yemen.
Correct orthography: *Ficus forskaolii* Vahl
Nomenclatural status: Illegitimate.
Currently used name for the taxon: ***Ficus palmata*** Forssk.

Papaveraceae

Mnemosilla Forskalii Ehrenb. ex Fedde, in Engl., *Pflanzenr.* 40 (IV.104) Papaverac.-Hypec.: 93, 1909. According to ICN, the eponym should be written *Mnemosilla forskalii* Ehrenb. ex Fedde. – It was proposed on a label without a description of the taxon and was later published in synonymy of *Hypecoum aegyptiacum* (Forssk.) Aschers. & Schweinf., in *Mém. Inst. Égypt.* 2: 37, 1887, based on *M. aegyptiaca* Forssk., *Fl. Aegypt.-Arab.*: LXIX, No. 314; 122, *Cent.* IV, No. 76, 1775, based on a plant from Egypt. As a name published in synonymy, it is not validly published.
Correct orthography: *Mnemosilla forskalii* Ehrenb. ex Fedde
Nomenclatural status: Not validly published.
Currently used name for the taxon: ***Hypecoum aegyptiacum*** (Forssk.) Aschers. & Schweinf.

Plantaginaceae (including Scrophulariaceae in Hepper & Friis 1994)

Simbuleta Forskåhlii J.F.Gmel., *Syst. Nat.*, ed. 13bis, 2(1): 242, 1791 [late Sep.–Nov. 1791]. The sign over the 'a' in the original publication is the lower part of an 'o' or an 'u' and the letter must be interpreted as an 'å,' the Scandinavian 'å' not being available in the typesetter's case (Fig. 10).
According to ICN, the orthography of the eponym must be corrected to *Simbuleta forskaohlii* J.F.Gmel. Cufodontis made a new combination in *Anarrhinum, A. forskaohlii* [as *Forskahlii*] (J.F.Gmel.) Cufod., in *Bull. Jard. Bot. État Bruxelles* 33 (Suppl.): 891, 1963. – In the protologue, Gmelin stated that he based his species name on a plant described by Forsskål, *Simbuleta* Forssk., *Fl. Aegypt.-Arab.*: 115, 1775, and the description was based on a plant from Yemen. Gmelin only cited the page with Arabic numbers, not specifically the *Cent.* IV, No. 54, here reproduced in Fig. 7, and not any mentioning in the floristic lists for Yemen with Roman numerals. In *Fl. Aegypt.-Arab.*: 115, there is a description of a new genus, named *Simbuleta* Forssk., with no specific epithet. In the floristic list with Roman numerals, on p. CXV, No. 389, here reproduced in Fig. 5, there is only the generic name with two asterisks, indicating that there is a description of a new genus elsewhere in the book. Thus, nowhere in *Fl. Aegypt.-Arab.*, Forsskål provided an epithet for a species name in his new genus *Simbuleta*. Gmelin, when providing a specific epithet, did not provide a separate description of the species. Cufodontis, when making the new combination in *Anarrhinum,* claimed that Forsskål's description of *Simbuleta* was a *descriptio generico-specifica* [which Cufodontis called a *diagnosis generico-specifica*], but without a validly published epithet for the species. However, Forsskål's protologue does not make up all requirements for a *descriptio generico-specifica* as defined in the ICN, Art. 38.5. Yet, according to ICN, Art. 38.12, Gmelin's species name, *S. forskaohlii,* is validated by reference to the previously published description of the new genus *Simbuleta*, and neither Forsskål, nor Gmelin indicated that more than one species belonged to the genus. Therefore, *Simbuleta forskaohlii* J.F.Gmel. is validly published and legitimate according to the ICN and a basionym for Cufodontis' *Anarrhinum forskaohlii* (J.F.Gmel.) Cufod.. Apparently unaware of Gmelin's work, Poiret published a new name for Forsskål's unnamed species, citing *Simbuleta* Forssk. *Fl. Aegypt.-Arab.* pag. 194, N°. 54, and naming the species *S. arabica* Poir., *Encycl. Meth. Bot.* 7: 194, 1806. Poiret's name is a superfluous, illegitimate renaming of *S. forskaohlii* J.F.Gmel.
Correct orthography: *Simbuleta forskaohlii* J.F.Gmel.
Nomenclatural status: Legitimate.
Currently used name for the taxon: ***Anarrhinum forskaohlii*** (J.F.Gmel.) Cufod.

Poaceae (Gramineae in Hepper & Friis 1994)

Anthistiria forskålii Kunth, *Révis. Gramin.* 1: 162, 1829.
According to ICN, the orthography of the eponym must be corrected to *Anthistiria forskaolii* Kunth. – In the protologue of his eponym, Kunth, *Révis. Gramin.* 1: 162, No. 20, 1829, cited *Themeda polygama* Forssk.

in synonymy. Forsskål never published that name, but it was proposed by Gmelin as *Themeda polygama* J.F.Gmel., *Syst. Nat.*, ed. 13bis. 2(1): 149, 1791, with a reference to Forsskål, *Fl. Aegypt.-Arab.*: 178, 1775. Forsskål's name for the plant on p. 178 is *Themeda triandra* Forssk., *Fl. Aegypt.-Arab.*: CXXIII, No. 598; 178, *Cent.* VI, No. 92, 1775, based on a plant from Yemen, but with no preserved material (Hepper & Friis 1994: 280). Gmelin's description begins with "Gramen polygama ...", just as Forsskål's description of *Themeda triandra*, and must be an indirect reference to Forsskål's name and the reason for Gmelin's and Kunth's errors. *Themeda triandra* Forssk. is legitimate, and both Gmelin's *T. polygama* and Kunth's *A. forskaolii* represent superfluous, illegitimate renamings.
Correct orthography: *Anthistiria forskaolii* Kunth
Nomenclatural status: Illegitimate.
Currently used name for the taxon: **Themeda triandra** Forssk. See also ***Themeda forskalii*** Hack.

Aristida Forskohlii Tausch, *Flora* 19(2): 506, 1836.
According to ICN, the eponym should be written as *Aristida forskohlii* Tausch. – In the protologue of his eponym, Tausch cited in synonymy Forsskål's '*A. tomentosa* Forsk. Descr. 25 (partim).' However, on p. 25 in Forssk., *Fl. Aegypt.-Arab.*, 1775, there is no *A. tomentosa*, only two species named *A. lanata* (from Egypt) and *A. paniculata* (from Yemen), which were both described as new by Forsskål. Tausch seems to have mixed two adjectives for hairiness, writing '*A. tomentosa* Forsk.' instead of *Aristida lanata* Forssk., *Fl. Aegypt.-Arab.*: LXI, No. 79; CIV, No. 79; 25, *Cent.* I, No. 87, 1775. Tausch claimed that Forsskål's *Aristida tomentosa* is based on a mixture of two species; for one he retained the name *Aristida tomentosa* and for the other he proposed the new name *A. forskohlii* Tausch. This complex situation was studied by Cope, who described it in a detailed note in Hepper & Friis (1994: 279). Cope concluded that all preserved Forsskål material associated with these names belongs to *Stipagrostis plumosa* (L.) Munro ex T.Anders., in *Journ. Linn. Soc. Bot.* 4, Suppl. 1: 40, 1860, and no Forsskål material agrees with the protologue of *A. lanata* Forssk., which is a legitimate name. We conclude that *A. forskohlii* Tausch is also legitimate because he has excluded Forsskål's *Aristida lanata* from it.
Correct orthography: *Aristida forskohlii* Tausch
Nomenclatural status: Legitimate.
Currently used name for the taxon: Without neotypification, it is not possible to identify *Aristida forskohlii* Tausch.

Avena Forskålei Vahl, *Symb. Bot.* 2: 25, 1791.
According to ICN, the orthography of the eponym must be corrected to *Avena forskaolii* Vahl. Brown and Trinius made new combinations with slight orthographic variants of the epithet in new combinations, *Danthonia forskaolii* [as *forskalii*] (Vahl) R.Br., in Denham & Clapperton, *Narr. Travels Africa*, Appendix: 91 (Boston ed.), and 244–245 (London ed.), 1826 [Mar.–Apr. 1826], and *D. forskaolii* [as *forskålii*] (Vahl) Trin., *Sp. Gram.* 1: 245, tab. 49, 1826. Nevski made a new combination, *Asthenatherum forskaolii* [as *forskolii*] (Vahl) Nevski, in *Trudy Sredne-Aziatsk. Gosud. Univ.*, Ser. 8b, Bot., 17: 8, 1934. More recently, Cope made the new combination *Centropodia forskaolii* [as *forskalii*] (Vahl) Cope, in *Kew Bull.* 37(4): 658, 1983 [20 Jul. 1983]. – In the protologue of his eponym, Vahl cited in synonymy *Avena pensylvanica* Forssk., *Fl. Aegypt.-Arab.*: LXI, No. 69; 23, *Cent.* I, No. 81, 1775, based on a plant from Egypt, but there was already an *A. pensylvanica* L., *Sp. Pl.* 1: 79, 1753, described from America. We conclude that Forsskål's name is a misidentification, *A. pensylvanica sensu* Forssk., non L. Therefore, the new name *A. forskaolii* Vahl is legitimate.
Correct orthography: *Avena forskaolii* Vahl
Nomenclatural status: Legitimate.
Currently used name for the taxon: ***Centropodia forskaolii*** (Vahl) Cope

Chaetaria Forskolii Nees, *Fl. Afr. Austral. Ill.*: 188, 1841.
According to ICN, Nees's eponym should be written as *Chaetaria forskolii* Nees. – In the protologue of his eponym, Nees cited in synonymy *Aristida paniculata* Forssk., *Fl. Aegypt.-Arab.*: CIV, No. 81; 25, *Cent.* I, No. 88, 1775, based on a plant from Yemen. Forsskål's name is legitimate, and *C. forskolii* Nees therefore rep-

resents a superfluous, illegitimate change of epithet. Both names are taxonomic synonyms of *A. adscensionis* L., *Sp. Pl.* 1: 82 (1753).
Correct orthography: *Chaetaria forskolii* Nees
Nomenclatural status: Illegitimate.
Currently used name for the taxon: ***Aristida adscensionis*** L.

Panicum Forskålii C.Chr., in *Dansk Bot. Ark.* 4(3): 12, 1922.
According to ICN, the orthography of the eponym must be corrected to *Panicum forskaolii* C.Chr. – In the protologue of his eponym, Christensen stated that *Panicum forskaolii* was a *nomen novum* for *Phalaris velutina* Forssk., *Fl. Aegypt.-Arab.*: 17, *Cent.* I, No. 55, 1775. However, Christensen did not cite the name that was published in the *Florulae* on p. CIII, No. 51; only the one in the *Descriptiones* on p. 17, *Cent.* I, No. 55. Forsskål's name was based on a plant from Yemen. Although Forsskål's name was legitimate, the species was renamed by Vahl as *Panicum sanguinale* Vahl, in *Symb. Bot.* 1: 8, 1790. However, this is an illegitimate later homonym of *P. sanguinale* L., *Sp. Pl.* 1: 57, 1753, now *Digitaria sanguinale* (L.) Scop., a species used as fodder and widely distributed as a weed. Thus, Christensen could neither use Forsskål's legitimate epithet, *velutina,* in *Panicum* because there was already a *P. velutinum* G.Mey., *Prim. Fl. Esseq.*: 51, 1818, nor Vahl's *sanguinale.* Christensen's name is a legitimate replacement name in *Panicum*, but the species is now transferred to *Digitaria*, under which there is no hindrance to take up Forsskål's original epithet, and the species is currently known as *D. velutina* (Forssk.) P.Beauv., *Ess. Agrostogr.*: 51 (1812).
Correct orthography: *Panicum forskaolii* C.Chr.
Nomenclatural status: Legitimate.
Currently used name for the taxon: ***Digitaria velutina*** (Forssk.) P.Beauv.

Themeda forskalii Hack., in A.DC. & C.DC., *Monogr. Phan.* 6: 659, 1889 [Apr. 1889].
And
A[nthistiria] Themeda Forskalii Duthie, *Fodder Grass. N. India*: 43, 1888.
According to ICN, the orthography of the eponym is correct. The year before Hackel's publication, Duthie published a variant of the name as *A.[nthistia] Themeda Forskalii* Duthie, in *Fodder Grasses N. India*: 43, 1888, but according to ICN, Art. 23.1, his name cannot be accepted because it is incorrectly formed. There are no references that connect Hackel's and Duthie's names. – In the protologue of his eponym, Hackel stated that he described a new species named after Forsskål because no authentic specimen was preserved, and he could therefore not decide to which of his varieties of *T. forskalii* Forsskål's name, *T. triandra* in the strict sense, belonged. The absence of any original material of *Themeda triandra* Forssk., *Fl. Aegypt.-Arab.*: CXXIII, No. 598; 178, *Cent.* VI, No. 92, 1775, is confirmed by Hepper & Friis (1994: 280), but Forsskål's name is legitimate and was based on a plant from Yemen, why Hackel's name is a superfluous, illegitimate renaming. The solution to Hackel's problem under the present ICN would be typification.
Correct orthography: *Themeda forskalii* Hack.
Nomenclatural status: Illegitimate.
Currently used name for the taxon: ***Themeda triandra*** Forssk.

Trisetaria Forskåhlii J.F.Gmel., *Syst. Nat.*, ed. 13bis, 2(1): 177, 1791 [late Sep.–Nov. 1791]. The sign over the 'a' in the original publication is part of an 'o' or an 'u' and the letter must be interpreted as an 'å,' the Scandinavian 'å' not being available in the typesetter's case (Fig. 10).
According to ICN, the orthography of the eponym must be corrected to *Trisetaria forskaohlii* J.F.Gmel. The similar name *Trisetum forskaolii* [as *Forskëlii*] P.Beauv., where we take the 'ë' to be a typographical error for 'å', has been listed in indexes, for example *Index Kewensis* (Jackson 1895b: 1124), as "*Trisetum Forskëlei* P.Beauv. = *Danthonia Forskålii*" [which is an unrelated species]. However, Palisot de Beauvais's name is not validly published because the new combination under *Trisetum*, based on *Trisetaria forskaohlii* J.F.Gmel., was not formally made. Palisot de Beauvois only recorded the epithet *Forskëlii* in a list of epithets which he intended to transfer to the genus *Trisetum*,

with which he considered Forsskål's *Trisetaria* to be congeneric. – In the protologue of his eponym, Gmelin referred to Forsskål, *Fl. Aegypt.-Arab.*: 27, *Cent.* I, No. 99, 1775, where Forsskål's new generic name *Trisetaria* is listed with a description based on a plant from Egypt, but without a specific epithet. However, a full name of the species is provided in the *Florulae*, which is *Trisetaria linearis* Forssk., *Fl. Aegypt.-Arab.*: LX, No. 52, 1775, and there is a double asterisk referring to the generic description, and together the two places in the book make up a *descriptio generico-specifica*. Therefore, the name *Trisetaria linearis* Forssk. is validly published, and *T. forskaohlii* J.F.Gmel. is a superfluous, illegitimate renaming.
Correct orthography: *Trisetaria forskaohlii* J.F.Gmel.
Nomenclatural status: Illegitimate.
Currently used name for the taxon: ***Trisetaria linearis*** Forssk.

Trit[icum] Forskal. Clemente, in Herrera, *Agric.* 1: 74, 1818.
According to ICN (including the accepted proposal at the Madrid nomenclatural session, McNeill *et al.* 2022), the orthography of the abbreviated eponym must be expanded in conformity with nomenclatural tradition. In this case, the only use of the name after its publication is in *Index Kewensis* (Jackson 1895b: 1127), where the abbreviated name is expanded to *Triticum forskalii* Clemente. According to ICN, Art. 60.14 and the accepted proposal in Madrid, the name in *Index Kewensis* represents a nomenclatural tradition, and the variant *Triticum forskalii* Clemente is correct. – Clemente did not cite a name by Forsskål in synonymy of his *T. forskalii*, but he stated that in Spanish the plant is called *Escanda vellosa*, which almost certainly must refer to Forsskål's *T. spelta* [var.] α *villosa* Forssk., *Fl. Aegypt.-Arab.*: LXI, No. 85; 26, *Cent.* I, No. 92, 1775, based on a plant from Egypt. In the brief description in Spanish, *Triticum forskalii* is said to differ by its villosity from *Trit[icum] spelta* L., *Sp. Pl.* 1: 86, 1753, or, as a subspecies, *T. aestivum* subsp. *spelta* (L.) Thell. in *Naturwiss. Wochenschr.*, n.s., 17: 471, 1918. Since there is no direct reference to names proposed by Forsskål, Clemente must be considered to have published a new name. The name has not been taken up for general use.
Correct orthography: *Triticum forskalii* Clemente
Nomenclatural status: Legitimate.
Currently used name for the taxon: ***Triticum spelta*** L. or ***Triticum aestivum*** L. subsp. ***spelta*** (L.) Thell.
Note: *Triticum Forskohlei* Lag. '(ex Trin[ius] mpt.)' is a name listed in Steudel, *Nomencl. Bot.*, ed. 2, 1: 716, 1840, but it has not been possible to clarify if the name is validly published or to establish its identity.

Ranunculaceae

Delphinium Forskålii Rchb., *Illustr. Sp. Acon.* 5: Tab. 68, 1823. The letter here interpreted as an 'å' is in the original publication an 'o' with a smaller 'o' above. There is no letter like than in the European alphabets we know of, but it must be treated in the same way as an identical letter in the designation of the author of '*Delphinium grandiflorum* Forskål,' where Forsskål's name is spelt with a small 'o' over the large second 'o,' obviously meaning 'å,' the Scandinavian 'å' not being available in the typesetter's case (Fig. 11).
According to ICN, the orthography of the eponym must be corrected to *Delphinium forskaolii* Rchb. – In the protologue of his eponym, Reichenbach cited in synonymy *D. grandiflorum* Forssk., *Fl. Aegypt.-Arab.*: XXVII, No. 251; 212, *Cent.* VIII, No. 56, 1775, based on a plant collected at the Dardanelles. However, there was already a *D. grandiflorum* L., *Sp. Pl.* 1: 531, 1753, described from Central Asia. We conclude that Forsskål's name is a misidentification, *D. grandiflorum sensu* Forssk., non L. *Delphinium forskaolii* Rchb. is therefore a legitimate new name, but is now considered a taxonomic synonym of *D. peregrinum* L., *Sp. Pl.* 1: 531, 1753.
Correct orthography: *Delphinium forskaolii* Rchb.
Nomenclatural status: Legitimate.
Currently used name for the taxon: ***Delphinium peregrinum*** L.

Ranunculus Forskæhlii DC., *Syst. Nat.* 1: 303, 1817 [publ. 1–15 Nov. 1817].
According to ICN, Art. 60.7, the orthography of the

eponym must be corrected to *Ranunculus forskoehlii* DC. because the ligature is composed of an 'o' and an 'e'. – When publishing his eponym, Candolle cited in synonymy *R. multifidus* Forssk., *Fl. Aegypt.-Arab.*: 102, *Cent.* IV, No. 18, 1775, based on a plant from Yemen. On p. 102, the new species is provided with Forsskål's name, *R. multifidus,* in full, with a description, and on p. CXIV, No. 350, 1775, the full species name is listed in the *Florulae*. It is not clear why Candolle chose to change this epithet, but the renaming is superfluous and illegitimate.
Correct orthography: *Ranunculus forskoehlii* DC.
Nomenclatural status: Illegitimate.
Currently used name for the taxon: ***Ranunculus multifidus*** Forssk.

Solanaceae

Solanum Forskalii Dunal, *Hist. Nat. Solanum*: 237, 1813 [19 Mar. 1813].
According to ICN, the eponym should be written *Solanum forskalii* Dunal. – In the protologue of his eponym, Dunal stated that he based *S. forskalii* on *S. villosum* Forssk., *Fl. Aegypt.-Arab.*: CVII, No. 151; 47, *Cent.* II, No. 57, 1775, a slender shrub collected by Forsskål in Yemen. In 1768, five years after Forsskål's death in Yemen and seven years before the publication of *Fl. Aegypt.-Arab.*, Miller had published *S. villosum* Mill., *Gard. Dict.*, ed. 8.: Species No. 2, 1768, an annual species grown from seeds collected on Barbados. Forsskål's name is therefore an illegitimate later homonym of Miller's name, and Dunal's name is a legitimate replacement name.
Correct orthography: *Solanum forskalii* Dunal
Nomenclatural status: Legitimate.
Currently used name for the taxon: ***Solanum forskalii*** Dunal

Urticaceae

Forsskålea L., *Opobalsamum*: 17, 1764.
Forsskålea tenacissima L., *Opobalsamum*: 19 [no p. 18; p. 19 follows immediately after p. 17], 1764.
According to ICN, the orthography of the eponym must be corrected to *Forsskaolea* L. (Fig. 9). – This is one of the cases where Forsskål's new species, and here also a new genus, has been described and named by Linnaeus from seeds sent by Forsskål to the Botanical Garden in Uppsala, but the genus and species had also been described and named by Forsskål himself in the field, in this case at Cairo in Egypt. For information about the consignments of seeds sent to a range of European botanical gardens from various places where the expedition made a halt, see Iamonico *et al.* (2015). In 1764, Linnaeus published *Forsskålea* L. and *F. tenacissima* L. in a dissertation with descriptions of both the genus and the species. It appears from the text of the dissertation that soon after Linnaeus had learnt about Forsskål's death in Yemen, he wrote words of remembrance for Forsskål and gave a genus the name *Forsskålea* dedicated to his memory. The names of the genus and species are legitimate. Forsskål's own names for the same plant, written in his fieldnotes, were only published eleven years later, in 1775, as *Caidbeja* Forssk., *Fl. Aegypt.-Arab.*: 82, Cent. III, No. 55, 1775, and *C. adhaerens* Forssk., *Fl. Aegypt.-Arab.*: LXV, No. 221; 82, *Cent.* III, No. 55, 1775. Forsskål's names for the genus and species are validated with a *descriptio generico-specifica* on p. 82 and the genus and species name appear on both p. LXV and on p. 82, where a reference with two asterisks indicates that there is a description of a new genus in the *Centuriae*. Also Forsskål's names of genus and species are legitimate.
Correct orthography: *Forsskaolea* L., *F. tenacissima* L.
Nomenclatural status: Legitimate.
Currently used name for the taxon: ***Forsskaolea*** L., ***F. tenacissima*** L.

Verbenaceae

Verbena Forskålæi Vahl, *Symb. Bot.* 3: 6, 1794.
According to ICN, the orthography of the eponym must be corrected to *Verbena forskaolii* Vahl. Meyer proposed a new combination, *Priva forskaolii* [as *Forskaolii*] (Vahl) E.Mey., *Comm. Pl. Afr. Austr.*: 275, 1835. Jaubert & Spach later proposed *P. forskaolii* [as *Forskalii*] (Vahl) Jaub. & Spach, *Pl. Orient.* 5(46): 59, tab.

455, 1855 [Jun. 1855]. – Vahl, in the protologue of *Verbena Forskålæi* Vahl in *Symb. Bot.*, cited the genus name *Phryma sensu* Forssk., *Fl. Aegypt.-Arab.*: CXV, No. 372 [here reproduced in Fig. 5], 1775, and based his *Verbena forskaolii* on a plant from Yemen with a new description. There is no reference in the protologue of *Verbena forskaolii* Vahl to Forsskål's *Ruellia adhaerens* Forssk. Vahl had a specimen collected by Forsskål in his own herbarium, now incorporated in the general collection at Herbarium C, and numbered *Forsskål 1814* [C10003245]. On the back of that sheet, Vahl had written 'Phryma? Forsk. flor. arab. no. 372 / Forskålei' (see Hepper & Friis 1994: 242). Thus interpreted, Vahl's name is a legitimate name for his new species, citing Forsskål's provisional identification to the genus *Phryma* in synonymy. However, Verdcourt (1988: 671), when revising the genus *Priva* for floras of eastern Africa, studied Forsskål's material from Yemen, noting that two rather similar species occurred in that country, clearly distinguishable on their mericarps, *Priva adhaerens* (Forssk.) Chiov. and a hitherto undescribed species, *P. tenax* Verdc. Verdcourt was certain that *Phryma?* Forssk. and Forsskål's *Ruellia adhaerens* Forssk., *Fl. Aegypt.-Arab.*: CXV, No. 388 [here reproduced in Fig. 5]; 114, *Cent.* IV, No. 51, 1775, were both based on the same material, both recorded from the same locality and both with the same indigenous name. For Forsskål's *Phryma,* this information is stated on p. CXV, No. 372: "*Hamsched.* Hadie. Musa. Seminibus 2 echinatis connatis. An *Ruellia.*" Moreover, for *Ruellia adhaerens,* this information is stated on p. CXV, no. 388: "*Hamsched.* Hadie." In the field, Forsskål obviously had access to all the plants from Hadie, and could very well, as suggested by Verdcourt, have based his notes on both *Phryma?* Forssk. and *Ruellia adhaerens* Forssk. on these plants. Only two collections made by Forsskål of these species exist in the Forsskål Herbarium at C: *Forsskål 1814* (C10003245, which is the specimen annotated by Vahl), and *Forsskål 1815* (C10003246, with no annotations by Vahl). Although without decisive evidence for our choice, we are inclined to follow Verdcourt's interpretation, to consider both specimens in the Forsskål Herbarium at C as original material of both *Ruellia adhaerens* Forssk. and *Verbena forskaolii* Vahl and to lectotypify both names with the best preserved specimen, *Forsskål 1814* (C10003245), although the formal typification is not made here. The currently accepted name, *Priva adhaerens* (Forssk.) Chiov., in *Bull. Soc. Bot. Ital.* 1923: 115, 1923, is based on *Ruellia adhaerens* Forssk.

Correct orthography: *Verbena forskaolii* Vahl

Nomenclatural status: Illegitimate (assuming that *Phryma?* Forssk. and *Ruellia adhaerens* Forssk. are based on the same type material).

Currently used name for the taxon: ***Priva adhaerens*** (Forssk.) Chiov.

Violaceae

Viola Forskaolii Greuter, in *Boissiera* 13: 59, 1967. According to ICN, the eponym should be written *Viola forskaolii* Greuter. – Although Greuter published the name in a footnote, all formal requirements for valid and legitimate publication are fulfilled. It is a new name for *Viola arborea* Forssk., *Fl. Aegypt.-Arab.*: CXX, No. 515, 1775, *nom. nud.*, ex Ging. & DC, in DC., *Prodr.* 1: 300, 1824, based on a plant from Yemen. There was already a *V. arborea* L., *Syst. Nat.*, ed. 10, 2: 1239, 1759, an American plant unrelated to Forsskål's. In the *Prodromus*, the plant is described as a new species, with a description based on a specimen with incomplete flowers seen by A.P. de Candolle in Herbarium Banks, '(DC. v.s. in h. Banks)', a *de facto* holotype (ICN, Art. 9.1. Note 1). The name validated in the *Prodromus* is an illegitimate later homonym of *V. arborea* L., as implied in the protologue of *Viola forskaolii* Greuter. Greuter did not cite any collecting number for the type material of his replacement name, referring to it as "holo- in herb. Banks, BM; iso- in herb. Forskål, C", and he mentioned only 'Yemen' as the place of collection, but the places of collection cited in *Fl. Aegypt.-Arab.* CXX, No. 515 ('Uahfud' or 'Uhf', now Wasab, and 'Barah', now Jabal Barad) can according to Hepper & Friis (1994: 243, Map in Fig. 9 on p. 59), and Wood (1997: 112) be localised to the slopes of the highlands of Yemen, and this agrees with Forsskål referring the collecting localities to his phytogeographical zone 'Mm', 'Montium regio media'. Hepper & Friis listed three

specimens of this species in the Forsskål Herbarium at C, but the holotype-specimen at BM cited by Greuter was not noted when Hepper searched the collections at BM for the account of Forsskål's plants (Hepper & Friis 1994). The specimen used by Gingins de la Sarraz & Candolle and by Greuter as type of *V. forskaolii* at BM has now the barcode BM013719705 and is annotated on the verso 'Ex Oriente Forskohl.' The isotypes at C are *Forsskål 621*, barcode C10003249; *Forsskål 622*, barcode C10003250; and *Forsskål 1823*, barcode C10003251. The material at BM and C can all be identified as *Hybanthus enneaspermus* (L.) F.Muell. (1876), now sometimes for taxonomic reasons transferred to the genus *Pigea* Ging. as *P. enneasperma* (L.) P.I.Forst., in *Austrobaileya* 11: 29, 2021. According to POWO (continuously updated) and TNRS (continuously updated), *V. arborea* Forssk. is a synonym of *V. stocksii* Boiss., *Fl. Orient.* 1: 453, 1867, but this is a mistake, probably caused by the fact that *V. stocksii* or *V. cinerea* Boiss. var. *stocksii* (Boiss.) W.Becker in *Beih. Bot. Centralbl.* 36(2): 37, 1918, is the only species of *Viola* known from Yemen. What seems to be overlooked is that *V. stocksii* is only recorded from a few places in southern Yemen, far from any locality visited by Forsskål. Therefore, *V. arborea* Ging. & DC. and *V forskaolii* Greuter are synonyms of *H. enneaspermus* (L.) F.Muell. or *P. enneasperma* (L.) P.I.Forst.
Correct orthography: *Viola forskaolii* Greuter
Nomenclatural status: Legitimate.
Currently used name for the taxon: ***Hybanthus enneaspermus*** (L.) F.Muell., or ***Pigea enneasperma*** (L.) P.I.Forst.

Vitaceae

Vitis Forskahlii Blatt., in *Rec. Bot. Surv. India* 8: 118, 1919.
According to ICN, the eponym should be written *Vitis forskahlii* Blatt. – In the protologue, Blatter cited in synonymy *Saelanthus rotundifolius* Forssk., *Fl. Aegypt.-Arab.*: CV, No. 100; 35, *Cent.* II, No. 14 & *Icon.*: Tab. 4, 1776, based on a plant from Yemen, as well as the homotypic synonyms *Cissus rotundifolia* Vahl, *Symb. Bot.* 3: 19, 1794, and *Vitis rotundifolia* Defl., *Voy. Yemen*: 125, 1889. See below about the valid publications of these names. For the latter, Deflers cited *Saelanthus rotundifolius* Forssk. and *Cissus rotundifolia* Vahl in synonymy. In his protologue, Blatter also cited Forsskål's and Vahl's names, as well as *Vitis rotundifolia* Defl. (1889), but pointed out that Deflers' name for the plant from Yemen was not identical with a species described earlier from America as *V. rotundifolia* Michx., *Fl. Bor.-Amer.* 2: 231, 1803. What neither Vahl, nor Deflers and Blatter would have known due to the rules of nomenclature in force at their times is that none of the names in Forsskål's new genus *Saelanthus* are validly published according to the current ICN, because there is no genus description for the four new species, a requirement in the ICN. Neither Vahl, nor Deflers and Blatter mentioned *Cissus rotundifolia* Lam., *Tabl. Encycl.* 1(2): 331, 1792. Lamarck cited in synonymy *S. rotundifolius* Forssk., and referred to the plate of Forsskål's plant in Forsskål (1776: Tab. 4), thus publishing the earliest legitimate name for this species. Vahl's name is an isonym, published later and independently of Lamarck's name. *Vitis forskahlii* Blatt. is not illegitimate: although he did not mention it, the legitimate *V. rotundifolia* Michx. (1803) prevented the transfer of the epithet *rotundifolia* from *Cissus rotundifolia* Lam. In *Cissus*, however, the correct name is *C. rotundifolia* Lam.
Correct orthography: *Vitis forskahlii* Blatt.
Nomenclatural status: Legitimate.
Currently used name for the taxon: ***Cissus rotundifolia*** Lam.

Zygophyllaceae

Fagonia Forskalii Pfund, in *Flora* 57: 414, 1874.
According to ICN, the eponym should be written *Fagonia forskalii* Pfund. – In the protologue of his eponym, Pfund claimed that it is a diagnostic feature of *F. forskalii* that the leaves of *F. arabica* L. and *F. cretica* L. fall before the fruits ripen, but that he had observed that this was not the case with the species of *Fagonia* that Forsskål had collected 'in desertis' at Cairo and called *F. scabra* Forssk., *Fl. Aegypt.-Arab.*: LXVI, No. 231; 88, *Cent.* III, No. 69, 1775. Pfund stated: "Der Forskål'sche

Name erscheint seiner mehrfachen Deutungen wegen unsicher; ich zog es daher vor, der Pflanze den Namen ihres ersten ungemein tätigen Entdecker beizulegen." [The identity of Forskål's name seems uncertain because of its possible multiple interpretations; I therefore preferred to give the plant the name of its first immensely active discoverer]. The problem stated by Pfund could have been solved by lectotypification of *Fagonia scabra* with one of the specimens in Forsskål's original material cited by Hepper & Friis (1994: 247). *Fagonia forskalii* Pfund is based on material collected by himself "in desertis pr. Cairo, Gebel achmer usque ad Atakka prope Suez." [in desert from Mt. Achmer near Cairo to Atakka near Suez.] Pfund's collections were deposited at B, and all material seems to have been destroyed. *Fagonia forskalii* Pfund is probably conspecific with *F. scabra* Forssk., but formal typification will be needed for certainty. *Fagonia scabra* Forssk. is now an accepted name for a species considered distinct from *F. cretica* L. Forsskål's species was transferred to the genus *Zygophyllum* in 2018 as *Z. scabrum* (Forssk.) Christenh. & Byng.

Correct orthography: *Fagonia forskalii* Pfund

Nomenclatural status: Legitimate.

Currently used name for the taxon: ***Fagonia scabra*** Forssk. or ***Zygophyllum scabrum*** (Forssk.) Christenh. & Byng.

Acknowledgements

The authors want to thank Rafaël Govaerts, Royal Botanic Gardens, Kew, for encouragement during this work with the orthography and other nomenclatural intricacies of Forsskål eponyms. We are also grateful to our two anonymous reviewers for working their way through an initial version of our complicated manuscript and offering us much detailed criticism and advice. Their many suggested alterations and corrections have led to a completely reworked text and generally been very useful, although pointing in completely opposite directions. One of the reviewers recommended only minimal orthographic corrections in relation to the orthography published in original publications, whereas the other advocated radical standardisation of the eponyms based on the spelling of Forsskål's name which we have established in this work.

We thank the Linnean Society of London for help with the letters from Forsskål to Linnaeus in their archives, and for scanning Forsskål's signature.

From earlier work with Forsskål's plants, Ib Friis remembers with gratitude both Frank Nigel Hepper (1929–2013), Kew, who involved his younger co-author deeply in the writing of the book *The Plants of Pehr Forsskål's* Flora Aegyptiaco-Arabica, and the members of the Special Committee on Binary Combinations (Attila Borhidi (Pécs), Charles Jeffrey (Kew), Lawrence Alexander Sidney Johnson (Sydney), Bengt Jonsell (Stockholm), and Hans Walter Lack (Berlin). The work of that committee resulted in changes in the ICN that made it possible to continue using Forsskål's *Flora Aegyptiaco-Arabica* as a source of validly published names, in spite of the shortcomings of that posthumously published work.

Literature cited

Anonymous [translation into Swedish by Laurentius Andreae, Olaus Petri and Laurentius Petri] (1541). *Biblia, thet är All then Heligha Scrifft på Swensko*. Uppsala. https://archive.org/details/SWEGVB_DBS_HS/mode/2up

Anonymous [Forsskål, P.] (1759). *Tankar om Borgerliga Friheten*. Stockholm: Laurentius Salvius. https://runeberg.org/forsborg/0001.html (and following pages to 0008.html).

Anonymous (1964–1966). Forsskål, Forsskåhl, Forskål. In: E. Grill (ed.), *Svenskt biografiskt lexikon*. Vol. 16: 358. Stockholm: Riksarkivet. https://sok.riksarkivet.se/sbl/artikel/14378

Anonymous (1968). Opinion 846. *Mullus auriflamma* Forsskål, 1775 (Pisces): suppressed under the plenary powers. *Bulletin of zoological Nomenclature* 25: 14–15. https://www.biodiversitylibrary.org/page/12223601#page/34/mode/1up

APG IV (2016). An update of the Angiosperm Phylogeny Group classification for the orders and families of flowering plants: APG IV. *Botanical Journal of the Linnean Society* 181: 1–20. https://doi.org/10.1111/boj.12385

Baack, L.J. (2013a). A naturalist of the Northern Enlightenment: Peter Forsskål after 250 years. *Archives of Natural History* 40(1): 1–19. https://www.euppublishing.com/doi/abs/10.3366/anh.2013.0132?widget=journalsociety

Baack, L.J. (2013b). From Biblical Philology to Scientific Achievement and Cultural Understanding. Carsten Niebuhr, Peter Forsskål and Frederik von Haven and the Transformation of the Danish Expedition to Arabia 1761–1767. [In Friis, I., Harbsmeier, M., & Simonsen, J.B. (eds.), *Early Scientific Expeditions and Local Encounters. New Perspectives on Carsten Niebuhr and 'The Arabian Journey'. Proceedings of a Symposium on the Occasion of the 250th Anniversary of the Royal Danish Expedition to Arabia Felix*]. Scientia Danica · Series H · Humanistica · 4.2: 61–77. Copenhagen: The Royal Danish Academy of Sciences and Letters. http://publ.royalacademy.dk/books/690/4948?lang=da

Baack, L.J. (2014). *Undying Curiosity. Carsten Niebuhr and the Royal Danish Expedition to Arabia (1761–1767)*. Stuttgart: Franz Steiner Verlag.

Bentham. G. (1844). Notes on the Mimoseae, with a synopsis of the species. Tribe III., Acacieae. *Hooker's London Journal of Botany* 3: 82–112. https://www.biodiversitylibrary.org/item/6314#page/86/mode/1up

Boulos, L. (1999). *Flora of Egypt*, Vol. 1 (Azollaceae-Oxalidaceae). Cairo: Al Hadara Publishing.

Boulos, L. (2000). *Flora of Egypt*, Vol. 2 (Geraniaceae-Boraginaceae). Cairo: Al Hadara Publishing.

Boulos, L. (2002). *Flora of Egypt*, Vol. 3 (Verbenaceae-Compositae). Cairo: Al Hadara Publishing.

Boulos, L. (2005). *Flora of Egypt*, Vol. 4. Monocotyledones (Alismataceae-Orchidaceae). Cairo: Al Hadara Publishing.

BPH (continuously updated). *BPH Online* [Database based on G.H.M. Lawrence *et al.* (editors), Botanico-Periodicum-Huntianum. 1968–1991]. Hunt Institute's Library. Pittsburgh: Hunt Institute for Botanical Documentation. https://www.huntbotanical.org/databases/show.php?1 (accessed 1 Jul. 2024).

Brongniart, A.T. (1825, as '1824'). Observations sur les végétaux fossiles renfermés dans les Grès de Hoer en Scanie. *Annales des Sciences Naturelles (Paris)* 4: 200–219. [The name is proposed on p. 210, illustrated on Planche 12, Fig. 3–6 (on p. 203).] https://bibdigital.rjb.csic.es/records/item/11390-annales-des-sciences-naturelles-tome-quatrieme

Briquet, J. (1906). *Règles internationales de la nomenclature botaniques adoptées par le Congrès International de Botanique de Vienne 1905*. Jena: Gustav Fischer. https://archive.org/details/rglesinternatiooobriqgoog

Briquet, J., Harms, H., Mangin, L. & Rendle, A.B. (1912). *Règles internationales de la nomenclature botaniques adoptées par le Congrès International de Botanique de Vienne 1905. Deuxième édition mise au point d'après les décisions du Congrès International de Botanique de Bruxelles 1910*. Jena: Gustav Fischer. https://gallica.bnf.fr/ark:/12148/bpt6k63827235

'British Botanists' (1954). Proposals for Modification of the International Code of Botanical Nomenclature. A. British Botanists. *Taxon* 3(2): 29–34. https://onlinelibrary.wiley.com/doi/epdf/10.2307/1217362 & https://doi.org/10.2307/1217362

Bruch, P., Schimper, W.P. & Gümbel, T. (1846). *Bryologia Europaea, seu Genera Muscorum Europaeorum*. Fasciculus XXXIII–XXXVI. Stuttgart: E. Schweizerbart. [The work appeared in fascicles, which were sold separately. The text on the genera were paginated separately. A facsimile edition with continuous pagination was published in 1971. The fascicles 33–36, which all appeared in 1846, were included on pp. 61–103 in volume 1 of the facsimile, in which *Angstroemia* is on pp. 169–172 & on Plate 94.]

Brummitt, R.K.& N.P. Taylor, N.P. (1990). To correct or not to correct? *Taxon* 39(2): 298–306. https://onlinelibrary.wiley.com/doi/epdf/10.2307/1223056 & https://doi.org/10.2307/1223056

Buhle, J.G. (1794). *Literarischer Briefwechsel von Johann David Michaelis. Erster Theil.* Leipzig: Weidmannschen Buchhandlung. https://archive.org/details/bim_eighteenth-century_literarischer-briefwechs_1794_1/mode/2up

Buhle, J.G. (1795). *Literarischer Briefwechsel von Johann David Michaelis. Zweyter Theil.* Leipzig: Weidmannschen Buchhandlung. https://books.google.dk/books?id=k8lbAAAAcAAJ&printsec=frontcover&hl=no&source=gbs_ge_summary_r&cad=0#v=onepage&q&f=false

Camp, W.H., Rickett, H.W. & Weatherby, C.A. (1947). International rules of botanical nomenclature: Formulated by the International Botanical Congresses of Vienna, 1905, Brussels, 1910, and Cambridge, 1930. Adopted and revised by the International Botanical Congress of Amsterdam, 1935. Unofficial special edition. *Brittonia* 6: 1–120. https://link.springer.com/article/10.2307/2804660

Candolle, A.P. de (1813). *Théorie élémentaire de la botanique, ou exposition des principes de la classification naturelle et de l'art de décrire et d'étudier les végétaux.* 1st ed. Paris: Déterville. https://www.biodiversitylibrary.org/bibliography/39705

Candolle, A.P. de (1819). *Théorie élémentaire de la botanique, ou exposition des principes de la classification naturelle et de l'art de décrire et d'étudier les végétaux.* 2nd ed., 'revue et augmentée.' Paris: Déterville. https://www.biodiversitylibrary.org/item/193841

Candolle, A. de (1867). *Lois de la nomenclature botanique.* Paris: V. Masson et fils. https://www.biodiversitylibrary.org/bibliography/139042

Carlsson, U. & Goldberg, D. (eds.) (2017). *The Legacy of Peter Forsskål. 250 Years of Freedom of Expression.* Göteborg: Nordicom. https://norden.diva-portal.org/smash/get/diva2:1534766/FULLTEXT01.pdf

Christensen, C. (1918). *Naturforskeren Pehr Forsskål – Hans Rejse til Ægypten og Arabien 1761–63 og hans botaniske Arbejder og Samlinger med 40 hidtil utrykte Breve og Dokumenter og et Portræt.* [The naturalist Pehr Forsskål – his journey to Egypt and Arabia and his botanical works and collections with 40 hitherto unpublished letters and documents and a portrait]. København: H. Hagerup. https://tekster.kb.dk/text/letters-002242618–000-root

Christensen, C. (1922). Index to Pehr Forsskål: Flora aegyptiaco-arabica 1775 with a Revision of Herbarium Forsskalii contained in The Botanical Museum of the University of Copenhagen. *Dansk Botanisk Arkiv* 4(3): 1–54.

Christensen, C. (1935). Forsskål, Petrus (Petter, Pehr), In: Engelstoft, P. & Dahl, S. (eds.), *Dansk Biografisk Leksikon.* Ed. 2, Vol. 7: 183–185. København: Gyldendal.

[Daniels, G.S. & Stafleu, F.A.] (1972). HI-IAPT Portraits of botanists, no. 11. Pehr Forsskal. *Taxon* 21(2–3): 252. https://onlinelibrary.wiley.com/doi/epdf/10.1002/j.1996–8175.1972.tb03259.x & https://doi.org/10.1002/j.1996–8175.1972.tb03259.x

Darbyshire, I., Kordofani, M., Farag, I., Candiga, R. & Pickering, H. (2015). *The Plants of Sudan and South Sudan* – An annotated checklist. Kew: Kew Publishing, Royal Botanic Gardens.

Demoulin, V. (2023). (414–425) Proposals on orthographical errors and orthographical variants. *Taxon* 72(5): 1183–1185. https://onlinelibrary.wiley.com/doi/epdf/10.1002/tax.13053 & https://doi.org/10.1002/tax.13053

Durazzini, A. (1772). Memoria sull'albero detto volgarmente Julibrissin, letta il dì 5 Agosto 1772, nella Società de' Georgofili dal Sig. Dott. Antonio Durazzini. *Magazzino Toscano* 3(4): 1–13 [?, not seen].

Edwards, S., Mesfin Tadesse & Hedberg, I. (1995). *Flora of Ethiopia and Eritrea*, Vol. 2(2) (Canellaceae to Euphorbiaceae). Addis Ababa, Ethiopia & Uppsala, Sweden.

Edwards, S., Sebsebe Demissew & Hedberg, I. (1997). *Flora of Ethiopia and Eritrea*, Vol. 6. (Hydrocharitaceae to Arecaceae). Addis Ababa, Ethiopia & Uppsala, Sweden.

Edwards, S., Mesfin Tadesse, Sebsebe Demissew, & Hedberg, I. (2000). *Flora of Ethiopia and Eritrea*, Vol. 2(1). (Magnoliaceae to Flacourtiaceae). Addis Ababa, Ethiopia & Uppsala, Sweden.

Fedosov, V., Fedorova, A., Ignatova, E. & Kučera, J. (2023). New Taxonomic Arrangement of *Dicranella* s.l. and *Aongstroemia* s.l. (Dicranidae, Bryophyta). *Plants (Basel)* 12(6): Article 1360: 38 pp. as PDF. Published online 2023, Mar. 17. Doi: 10.3390/plants12061360

Fleischer, M. (1904, as 1900–1902). *Die Musci der Flora Von Buitenzorg (Zugleich Laubmoosflora Von Java)*. Vol. 1. Leiden: E.J. Brill. https://www.biodiversitylibrary.org/item/98330

Forsskåhl, G. 1927. *Några blad ur släkten Forsskåhls historia, jämte några data angående dess kvinnosidas linjer: Biografiska anteckningar 1556–1927*. Borgå.

Forsskål, P. (1775a). *Flora Aegyptiaco-Arabica. Sive descriptiones plantarum quas per Aegyptum inferiorem et Arabiam Felicem detexit, illustravit Petrus Forskål. Post mortem auctoris edidit Carsten Niebuhr.* Copenhagen: Møller. https://www.biodiversitylibrary.org/bibliography/6625

Forsskål, P. (1775b). *Descriptiones animalium: avium, amphibiorum, piscium, insectorum, vermium; quae in itinere orientali observavit Petrus Forskål. Post mortem auctoris edidit Carsten Niebuhr.* Copenhagen: Møller. https://www.biodiversitylibrary.org/bibliography/2154

Forsskål, P. (1776). *Icones rerum naturalium, quas in itinere orientali depingi curavit Petrus Forsskål. Post mortem auctoris ad regis mandatum aeri incisas edidit Carsten Niebuhr.* Copenhagen: Møller. https://www.biodiversitylibrary.org/bibliography/70772

Fox Maule, A. (1984). Martin Vahl and his Symbolae. Botanicae. [With translation of Martin Vahl's Latin foreword to Vol. 1 of *Symbolae*]. Pp. xi–xiv in Vahl, M., Symbolae botanicae *or exact descriptions of plants collected by Petrus Forsskål on his Arabian journey and other recently discovered plants together with observations of some previously known plants.* Part 1. (Faximile edition). Copenhagen: Scandinavian Fine Editions.

Fricke, R. (2008). Authorship, availability and validity of fish names described by Peter (Pehr) Simon Forsskål and Jonas Christian Fabricius in the 'Descriptiones animalium' by Carsten Niebuhr in 1775 (Pisces). *Stuttgarter Beiträge zur Naturkunde A, Neue Series*, 1: 1–76. https://www.zobodat.at/pdf/Stuttgarter-Beitraege-Naturkunde_NS_1_A_0001-0076.pdf

Fries, T.M. (1903). *Linné. Lefnadsteckning.* Senare delen. Stockholm: Fahlcrantz & Co.

Fries, T.M. (1912). *Bref och Skrifvelser af och till Carl von Linné. Första Afdelningen Del VI.* Bref till och från svenska enskilda personer: E–Hallman. Stockholm: Aktiebolaget Ljus. https://archive.org/details/brefochskrifvels06uppsuoft/page/144/mode/2up?view=theater

Friis, I. (1984). An analysis of the protologues of new genera with new species in Forsskål's *Flora Aegyptiaco-Arabica*. *Taxon* 33(4): 655–667. https://onlinelibrary.wiley.com/doi/epdf/10.2307/1220782 & https://doi.org/10.2307/1220782

Friis, I. (1992). Proposals to amend the ICBN (63–73). Ten proposals to amend the Code, and report of the Special Committee on Binary Combinations. *Taxon* 41(2): 343–350. https://onlinelibrary.wiley.com/doi/epdf/10.2307/1222349 & https://doi.org/10.2307/1222349

Friis, I. & Thulin, M. (1984). The spelling of Pehr Forsskål's family name. *Taxon* 33(4): 668–672. https://onlinelibrary.wiley.com/doi/epdf/10.2307/1220783 & https://doi.org/10.2307/1220783

Friis, I. & Jeffrey, C. (1986). Proposal to revise ICBN (93). The problem of non-binary species names in works published after 1 May 1753, with a proposal to amend ICBN. *Taxon* 35(2): 397–400. https://onlinelibrary.wiley.com/doi/epdf/10.1002/j.1996-8175.1986.tb02548.x & https://doi.org/10.1002/j.1996-8175.1986.tb02548.x

Friis, I., Brummitt, R.K., Hepper, F.N., Jeffrey, C., Taylor, N.P., Verdcourt, B., Howard, R.A. & Nicolson, D.H. (1984). Validity of names published by Forsskål and Aublet. *Taxon* 33(3): 495–496. https://onlinelibrary.wiley.com/doi/epdf/10.1002/j.1996-8175.1984.tb03904.x & https://doi.org/10.1002/j.1996-8175.1984.tb03904.x

Friis, I. & Thulin, M. (2023). (3001) Proposal to conserve the name *Jatropha pelargoniifolia* against *Adenoropium forskaolii* (*Euphorbiaceae*). *Taxon* 72(6): 1373–1374. https://onlinelibrary.wiley.com/doi/epdf/10.1002/tax.13094 & https://doi.org/10.1002/tax.13094

GBIF (continuously updated). *Global Biodiversity Information Facility*. https://www.gbif.org/ (accessed 1 Jul. 2024).

Global Plants (continuously updated). Global Plants on JSTOR. https://plants.jstor.org/ (accessed 1 Jul. 2024).

Govaerts, R., Nic Lughadha, E., Black, N., Turner, R. & Paton, A. (2021). The World Checklist of Vascular Plants, a continuously updated resource for exploring global plant diversity. *Nature Scientific Data* 8: Article 215 (10 pp.). https://doi.org/10.1038/s41597–021–00997–6

Greuter, W. 1984. Warning against misinterpreting the rule on 'non-Linnean' works (Art.23.6(c). *Taxon* 33(3): 493–495. https://onlinelibrary.wiley.com/doi/epdf/10.1002/j.1996–8175.1984.tb03903.x & https://doi.org/10.1002/j.1996–8175.1984.tb03903.x

Greuter, W., Brummitt, R.K., Farr, E., Kilian, N., Kirk, P.M. & Silva, P.C. (1993). NCU-3. Names in Current Use for Extant Plant Genera. *Regnum Vegetabile* 129: i–xxvii & 1–1464.

Greuter, W., Burdet, H.M., Chaloner, W.G., Demoulin, V., Grolle, R., Hawksworth, D.L., Nicolson, D.H., Silva, P.C., Stafleu, F.A., Voss, E.G. & McNeill, J. (1988). International Code of Botanical Nomenclature. Adopted by the Fourteenth International Botanical Congress, Berlin, July–August 1987. *Regnum Vegetabile* 118: i–xiv &1–328. https://archive.org/details/internationalcod88inte

Greuter, W., Barrie, F.R., Burdet, H.M., Chaloner, W.G., Demoulin, V., Hawksworth, D.L., Jørgensen, P.M., Nicolson, D.H., Silva, P.C., Trehane, P. & McNeill, J. (1994). International Code of Botanical Nomenclature (Tokyo Code), Adopted by the Fifteenth International Botanical Congress, Yokohama, August–September 1993. *Regnum Vegetabile* 131: i–xviii & 1–389. https://www.bgbm.org/iapt/nomenclature/code/tokyo-e/

Guedes, P., Alves-Martins, F., Martínez Arribas, J., Chatterjee, S., Santos, Ana M.C., Lewin, A., Bako, L., W. Webala, P.W., Correia, R.A., Rocha, R. & Ladle, R.J. (2023). Eponyms have no place in 21st century biological nomenclature. *Nature Ecology & Evolution* [Digitally published]. https://doi.org/10.1038/s41559–023–02022-y

Haeckel, E. (1888). Report on the Siphonophorae collected by HMS Challenger during the years 1873–1876. In: *Report of the Scientific Results of the Voyage of H.M.S. Challenger*, Zoology, Vol. 28: 1–380. https://www.biodiversitylibrary.org/item/18411#page/384/mode/1up

Harms, H., Rendle, A.B. & Hochreutiner, B.P.G. (eds.) (1935). *International Rules of Botanical Nomenclature, adopted by the international botanical congresses of Vienna, 1905, and Brussels, 1910, revised by the International Botanical Congress of Cambridge, 1930, compiled by the editorial committee for nomenclature from the report of the subsection of nomenclature.* Jena: Gustav Fischer.

Haugen, E. (1976). *The Scandinavian Languages. An Introduction to Their History*. Cambridge, Massachusetts: Harvard University Press. https://dn790003.ca.archive.org/0/items/haugen-the-scandinavian-languages-1976/Haugen%20-%20The%20Scandinavian%20Languages%20%281976%29.pdf

Hedberg, I. & Edwards, S. (1989). *Flora of Ethiopia*, Vol. 3. (Pittosporaceae to Araliaceae). Addis Ababa and Asmara, Ethiopia & Uppsala, Sweden.

Hedberg, I. & Edwards, S. (1995). *Flora of Ethiopia*, Vol. 7. (Poaceae (Gramineae), by Phillips, S.). Addis Ababa, Ethiopia & Uppsala, Sweden.

Hedberg, I., Edwards, S. & Sileshi Nemomissa (2003). *Flora of Ethiopia and Eritrea*, Vol. 4(1). (Apiaceae to Dipsacaceae). Addis Ababa, Ethiopia & Uppsala, Sweden.

Hedberg, I., Friis, I. & Edwards, S. (2004). *Flora of Ethiopia and Eritrea*, Vol. 4(2). (Asteraceae (Compositae)). Addis Ababa, Ethiopia & Uppsala, Sweden.

Hedberg, I., Ensermu Kelbessa, Edwards, S., Sebsebe Demissew & Person, E. (2006). *Flora of Ethiopia and Eritrea*, Vol. 5. (Gentianaceae to Cyclocheilaceae). Addis Ababa, Ethiopia & Uppsala, Sweden.

Hedberg, I., Friis, I. & Persson, E. (2009a). *Flora of Ethiopia and Eritrea*, Vol. 1. (Lycopodiaceae to Pinaceae; Appendix). Addis Ababa, Ethiopia & Uppsala, Sweden.

Hedberg, I., Friis, I. & Persson, E. (2009b). *Flora of Ethiopia and Eritrea*, Vol. 8. (General part and Index to Vol. 1–7). Addis Ababa, Ethiopia & Uppsala, Sweden.

Hepper, F.N. & Friis, I. (1994). *The Plants of Pehr Forsskål's* Flora Aegyptiaco-Arabica. Kew: Royal Botanic Gardens, & Copenhagen: Botanical Museum.

Herendeen, P.S. (2015). Report of the Nomenclature Committee on Fossils: 9. *Taxon* 64(6): 1306–1312. https://onlinelibrary.wiley.com/doi/epdf/10.12705/646.14 & http://dx.doi.org/10.12705/646.14

Hildebrand, B. (1964–1966). Forsskål, Johannes. In: E. Grill (ed.), *Svenskt biografiskt lexikon.* Vol. 16: 358. Stockholm: Riksarkivet. https://sok.riksarkivet.se/sbl/artikel/14379

Hildebrand, B. & Matinolli, E. (1964–1966). Forsskål, Peter (Petrus). In: E. Grill (ed.), *Svenskt biografiskt lexikon.* Vol. 16: 359. Stockholm: Riksarkivet. https://sok.riksarkivet.se/sbl/artikel/14380

Hylander, N. (1968). Various proposals. Note on proposal 116. *Taxon* 17(5): 592–593. https://onlinelibrary.

wiley.com/doi/epdf/10.2307/1216078 & https://doi.org/10.2307/1216078

Iamonico, D., Friis, I., Jarvis, C.E. & Freitag, H. (2015). Nomenclatural note on the genus *Pteranthus* (Caryophyllaceae) with lectotypification of the name *Camphorosma pteranthus* and *Pteranthus trigynus*, and comments on Forsskål's seeds sent from Egypt. *Taxon* 64(4): 1–6. https://onlinelibrary.wiley.com/doi/epdf/10.12705/644.13 & https://doi.org/10.12705/644.13

International Commission on Zoological Nomenclature. (1999). *International Code of Zoological Nomenclature (ICZN)*. Fourth Edition. https://code.iczn.org/ (accessed 12 Jun. 2024).

IPNI (continuously updated). *The International Plant Name Index*. http://www.ipni.org/ (accessed 1 Jul. 2024).

Jackson, B.D. (1895a). *Index Kewensis. Plantarum phanerogamarum nomina et synonyma* ... Vol. 1 [separate pagination]. Oxford: Clarendon Press. https://www.biodiversitylibrary.org/bibliography/66720

Jackson, B.D. (1895b). *Index Kewensis. Plantarum phanerogamarum nomina et synonyma* ... Vol. 2 [separate pagination]. Oxford: Clarendon Press. https://www.biodiversitylibrary.org/bibliography/66

Janchen, E. (1954). Anträge von Erwin Janchen (Wien). *Taxon* 3(2): 34–39. https://onlinelibrary.wiley.com/doi/epdf/10.1002/j.1996–8175.1954.tb01537.x & https://doi.org/10.1002/j.1996–8175.1954.tb01537.x

Jeffrey, C. (1985). Forsskål and the interpretation of article 23. *Taxon* 34(1): 144–147. https://onlinelibrary.wiley.com/doi/epdf/10.2307/1221579 & https://doi.org/10.2307/1221579

Knapp, S., Alvarez, I., Monro, A., Prado, J., Turland, N.J. & Wiersema, J. (2024). Livestreaming the Nomenclature Section at the XX International Botanical Congress in Madrid. *Taxon* 73(1): 413–414. https://onlinelibrary.wiley.com/doi/epdf/10.1002/tax.13127 & https://doi.org/10.1002/tax.13127

Kölliker, A. (1853). *Die Schwimmpolypen oder Siphonophoren von Messina*. Leipzig: Wilhelm Engelmann. https://www.biodiversitylibrary.org/page/12726286#page/9/mode/1up

Lagus, W. (1877). *Strödda blad, nytt och gammalt i historiska och humanistiska ämnen, I. Petter Forskåls lefnad*. Helsingfors: G.W. Edlund.

Lanjouw, J. (1954). Recueil synoptique des propositions concernant le Code international de la nomenclature botanique soumises à la Section de nomenclature du 8. Congrès international de Botanique, Paris, 1954. *Regnum Vegetabile* 4: 1–124.

Lanjouw, J., Baehni, C., Merrill, E.D., Rickett, H.W., Robyns, W., Sprague, T.A. & Stafleu, F.A. (1952). International Code of Botanical Nomenclature. Adopted by the Seventh International Botanical Congress, Stockholm, July 1950. *Regnum Vegetabile* 3: 1–228.

Lanjouw, J., Baehni, C., Robyns, W., Rollins, R.C., Ross, R., Rousseau, J., Schulze, G.M., Smith, A.C., De Vilmorin, R. & Stafleu, F.A. (1956). International Code of Botanical Nomenclature. Adopted by the Eighth International Botanical Congress, Paris, July 1854. *Regnum Vegetabile* 8: 1–338.

Lanjouw, J., Baehni, C., Robyns, W., Ross, R., Rousseau, J., Schopf, J.M., Schulze, G.M., Smith, A.C., De Vilmorin, R. & Stafleu, F.A. (1961). International Code of Botanical Nomenclature. Adopted by the Ninth International Botanical Congress, Montreal, August 1959. *Regnum Vegetabile* 23: 1–372.

Linnaeus, C. (1751). *Philosophia botanica*. Stockholm (Holmiae): Kiesewetter. https://www.biodiversitylibrary.org/item/84231

Linnaeus, C. (1758). *Systema naturæ, per regna tria naturæ; secundum classes, ordines, genera, species, cum characteribus, differentiis, synonymis, locis*. Tomus I. Stockholm (Holmiae): Laurentius Salvius. https://archive.org/details/mobot31753000798865/page/n1/mode/2up

Linnaeus, C. (1764). *Opobalsamum declaratum in dissertatione medica, cum consensu exper. Facultat. medicae, sub praesidio ... Caroli von Linné, ... publico examini modeste submittit ... Wilhelmus Le Moine*, ... Upsaliae. https://uu.diva-portal.org/smash/get/diva2:1378385/FULLTEXT01.pdf

Little, E.L. (1945). Miscellaneous Notes on Nomenclature of United States Trees. *The American Midland Naturalist* 33(2): 495–513. DOI. https://www.jstor.org/stable/2421343

Lohmeier, D. (1994). Richolff, Jürgen (Georg) d. J. In: *Biographisches Lexikon für Schleswig-Holstein und Lübeck*, Vol. 10. Kiel/Hamburg, Wachholtz Verlag. Here cited from p. 2265 in the digital edition by Schleswig-Holsteinische Landesbibliothek (https://files.wachholtz-verlag.de/openaccess/9783529025624.pdf)

Mabberley, D.J. (1978). Edward and Sarah Bowdich's Names of Macaronesian and African Plants, with Notes on those of Robert Brown. *Botánica Maca-*

ronésica 6: 53–66. https://dialnet.unirioja.es/servlet/articulo?codigo=2688343

Mabberley, D.J. & Moore, D.T. (2022). The Robert Brown Handbook. *Regnum Vegetabile* 160: 1–623.

Margadant, W.D. & Geissler, P. (1995). (1175–1191) Seventeen proposals concerning nomina conservanda for genera of Musci. *Taxon* 44(4): 613–624. https://onlinelibrary.wiley.com/doi/epdf/10.2307/1223506 & https://doi.org/10.2307/1223506

McNeill, J., Greuter, W. & Hartley, H. (2022). (186–187) Proposals to add a new provision in Article 60 to establish that epithets formed from abbreviations of personal names are considered to be composed arbitrarily and the termination of such epithets is not subject to standardisation, and to clarify Article 60. *Taxon* 71(6): 1340–1341. https://onlinelibrary.wiley.com/doi/epdf/10.1002/tax.12852 & https://doi.org/10.1002/tax.12852

Nicolson, D.H. (1981). Report of the Special Committee on Romanization of authors' names from non-Roman scripts. *Taxon* 30(1): 168–183. https://onlinelibrary.wiley.com/doi/epdf/10.1002/j.1996–8175.1981.tb00734.x & https://doi.org/10.1002/j.1996–8175.1981.tb00734.x

Nicolson, D.H. (1999). Report of the General Committee: 8. *Taxon* 48(2): 373–378. https://onlinelibrary.wiley.com/doi/epdf/10.2307/1224444 & https://doi.org/10.2307/1224444

Niebuhr, B.G. (1816). Carsten Niebuhrs Leben. [The life of Carsten Niebuhr]. *Kieler Blätter* 3: 1–86. https://books.google.dk/books?id=FkdRAAAAcAAJ

Niebuhr, B.G. (1817). Carsten Niebuhrs Levnet. [The life of Carsten Niebuhr]. *Athene, et Maaneds Skrift* [Copenhagen] 8: 1–36 & 117–166.

Niebuhr, C. (1772). *Beschreibung von Arabien.* Aus eigenen Beobachtungen und im Lande selbst gesammelten Nachrichten abgefasst. [Description of Arabia. Based on own observations and information gathered by the author in the countries]. With 24 plates and 1 map. Kopenhagen: Møller; Leipzig: Breitkopf und Sohn. https://wellcomecollection.org/works/f9gean9n/items

Niebuhr, C. (1774–1778). *Reisebeschreibung nach Arabien und anderen umliegenden Ländern.* [Description of travel to Arabia and other surrounding countries]. Vol. 1–2, with 124 plates and 1 map. Copenhagen: Møller. https://www.digitale-sammlungen.de/de/view/bsb11122255?page=1 & https://www.digitale-sammlungen.de/de/view/bsb11122256?q=%28Niebuhr+Reisebeschreibung+Zweiter+Band%29&page=6,7

Noreen, A. (1913). Dekorativ stavning av familjenamn. Pp. 122–131 in Noreen, A. *Spridda studier. Tredje samlingen. Populära uppsatser.* Stockholm: Ljus. DOI: https://runeberg.org/naspristud/3/0135.html

Notaris, G. de (1869). Epilogo della briologia Italiana. *Atti della Reale Università Genova* 1: i–xxiv & 1–781. https://bibdigital.rjb.csic.es/records/item/10211-epilogo-della-briologia-italiana

Paclt, J. (1968). Nomenclature proposals to the Seattle Congress. *Taxon* 17(3): 333–334.

Paris, É.G. (1894). Index Bryologicus (part). *Actis Societatis Linnaeanae Burdigalensis* Sér. 5, Vol. 6, Part 1: 15–62.

Pedersen, S.S. (2004). Carsten Niebuhr som forlægger. [Carsten Niebuhr as publisher]. *Magasin fra Det Kongelige Bibliotek* 17(4): 51–56. https://tidsskrift.dk/magasin/article/view/66570/95907

POWO (continuously updated). *Plants of the World Online.* Kew: Royal Botanic Gardens. https://powo.science.kew.org/ (accessed 1 Jul. 2024).

Provençal, P. (2010). The Arabic Plant Names of Peter Forsskål's *Flora Aegyptiaco-Arabica*. *Biologiske Skrifter* 57: 1–155. https://www.royalacademy.dk/Publications/Low/2824_Proven%C3%A7al,%20Philippe.pdf

Pugh, P.R. (2003) A revision of the family Forskaliidae (Siphonophora, Physonectae), *Journal of Natural History* 37(11): 1281–1327. DOI: 10.1080/00222930110120638

Rasmussen, S.T. (ed.) (1990). *Den Arabiske Rejse 1761–1767. En dansk ekspedition set i videnskabshistorisk perspektiv.* [The Arabian Voyage 1761–1767. A Danish Expedition seen through the History of the Sciences and Humanities.] København: Munksgaard.

Regnéll, G. (1952). Wiedergabe des skandinavischen å durch das Klassische Latein-Alphabet. *Senckenbergiana* 33(1–3): 69–71.

Rendle, A.B. (1934). International Rules of Botanical Nomenclature adopted by the Fifth International Botanical Congress, Cambridge, 1930. *London Journal of Botany* 72: 858 & Suppl. 1–29.

Rijckevorsel, P. van (2023). (228–232) Proposals on orthography: Diacritical signs. *Taxon* 72(2): 458–459. DOI https://doi.org/10.1002/tax.12912

Rousseau, J., Baehni, C., Rickett, H.W., Robyns, W., Lanjouw, J., Stafleu, F.A. & Pichon, M. (1955). Huitieme Congres International de Botanique. Section nomenclature. Suite des discussions. *Taxon* 4(7): 153–177.

Sandersen, V. (2002). Om bogstavet å. *Nyt fra Sprognævnet* 2002(2): Unpaginated, pp. [1]–[7]. https://dsn.dk/wp-content/uploads/2021/01/september-2002-pdf.pdf

Schweinfurth, G. (1889). Über seine Reise nach dem glücklichen Arabien. *Verhandlungen der Gesellschaft für Erdkunde zu Berlin* 16: 299–308.

Schweinfurth, G. (1891). Über die Florengemeinschaft von Südarabien und Nordabessinien. *Verhandlungen der Gesellschaft für Erdkunde zu Berlin* 18: 531–550.

Schweinfurth, G. (1912). *Arabische Pflanzennamen aus Aegypten, Algerien und Yemen*. Berlin: Dietrich Reimer (Ernst Vohsen).

Sherborn, C.D. (1902). *Index animalium sive index nominum quae ab A.D. MDCCLVIII generibus et speciebus animalium imposita sunt. Sectio prima*. A Kalendis Ianuariis, MDCCLVIII usque ad finem Decembris, MDCCC. Cantabrigiae: Typographico Academico.

Sherborn, C.D. (1922–1932). *Index animalium sive index nominum quae ab A.D. MDCCLVIII generibus et speciebus animalium imposita sunt*. Sectio secunda. A Kalendis Ianuariis, MDCCCI usque ad finem Decembris, MDCCCL. London: Longmans, Green & Co. & British Museum (Natural History)

Schück, H. (1923). *Från Linnés tid. Petter Forsskål*. Stockholm: P.A. Norstedt & Söners Förlag.

Skautrup, P. 1953. Å-ets historie. In: V. Dybdahl, B. Jensen & E. Sejr (eds), *Seksten Århusrids. – tilegnede S. Unmack Larsen, 25. Sept. 1953*: 291–315. Aarhus: Universitetsforlaget i Aarhus.

Stafleu, F.A., Bonner, C.E.B., McVaugh, R., Meikle, R.D., Rollins, R.C., Ross, R., Schopf, J.M., Schulze, G.M., De Vilmorin, R., & Voss, E.G. (1972). *International Code of Botanical Nomenclature, adopted by the Eleventh International Botanical Congress, Seattle, August 1969*. Utrecht: A. Osthoek's Uitgeversmaatschappij. 426 pp.

Stafleu, F.A. & Cowan, R.S. (1976). Taxonomic literature. Ed. 2. Vol. 1.: A–G. *Regnum Vegetabile* 94: i–xl & 1–1136. https://www.biodiversitylibrary.org/bibliography/48631

Stearn W.T. (1961). Introductory notes on Linnaeus's 'Mantissa Plantarum'. Pp. iii–xxv in Linnaeus, C., *Mantissa Plantarum* 1767 and 1771 (Facsimile edition). Weinheim: J. Cramer.

Stearn W.T. (1973). *Botanical Latin. History, Grammar, Syntax, Terminology and Vocabulary*. Second Edition. Newton Abbot: David & Charles.

Targioni-Tozzetti, A. (1853). *Cenni storici sulla introduzione di varie piante nell' agricoltura ed orticoltura Toscana*. Firenze: Tipografia Galiletana. https://www.biodiversitylibrary.org/item/48254#page/5/mode/1up

Thiers, B.M. (continuously updated). *Index Herbariorum*. New York, Steere Herbarium, New York Botanical Garden's Virtual Herbarium. Available at https://sweetgum.nybg.org/science/ih/ (accessed 28 Jun. 2024).

Thulin, M. (1993). *Flora of Somalia*, Vol. 1. Pteridophyta; Gymnospermae; Angiospermae (Annonaceae–Fabaceae). Kew: Royal Botanic Gardens.

Thulin, M. (1995). *Flora of Somalia*, Vol. 4. Angiospermae (Hydrocharitaceae–Pandanaceae) Kew: Royal Botanic Gardens.

Thulin, M. (1999). *Flora of Somalia*, Vol. 2. Angiospermae (Tiliaceae–Apiaceae). Kew: Royal Botanic Gardens.

Thulin, M. (2006). *Flora of Somalia*, Vol. 3. Angiospermae (cont.). Kew: Royal Botanic Gardens.

Thulin, M. (2016). Disentangling the history of Forsskål's 'Camellia.' *Symbolae Botanicae Upsalienses* 38: 139–143.

TL-2 (2024). *Smithsonian Libraries: Taxonomic Literature II (TL-2)*. [Online edition]. https://www.sil.si.edu/DigitalCollections/tl-2/index.cfm (accessed 1 Jul. 2024)

TNRS (continuously updated). *Taxonomic Name Resolution Service*. https://tnrs.biendata.org/ (accessed 1 Jul. 2024).

Turland, N.J., Wiersema, J.H., Barrie, F.R., Greuter, W., Hawksworth, D.L., Herendeen, P.S., Knapp, S., Kusber, W.-H., Li, D.-Z., Marhold, K., May, T.W., McNeill, J., Monro, A.M., Prado, J., Price, M.J. & Smith, G.F. (eds.) (2018). International Code of Nomenclature for Algae, Fungi, and Plants (*Shenzhen Code*) adopted by the Nineteenth International Botanical Congress Shenzhen, China, July 2017. [= ICN]. *Regnum Vegetabile* 159: i–xxxviii & 1–254. DOI https://doi.org/10.12705/Code.2018

Turland, N.J. & Wiersema, J.H. (2024). Synopsis of Proposals on Nomenclature – Madrid 2024: A review of the proposals to amend the International Code of Nomenclature for algae, fungi, and plants submitted to the XX International Botanical Congress. *Taxon* 73(1): 325–404. DOI https://doi.org/10.1002/tax.1311

Turland, N.J., Alvarez, I, Knapp, S., Monro, A.M. & Wiersema, J.H. (2024). XX International Botanical

Congress, Madrid 2024: Report of Congress action on nomenclature proposals. *Taxon* 73(5): 1308–1323. DOI https://doi.org/10.1002/tax.13258
Vahl, M. (1790). *Symbolae botanicae, sive plantarum, tam earum, quas in itinere, inprimis orientali, collegit Petrus Forskål,* ... Vol. 1. Copenhagen: Møller. https://www.biodiversitylibrary.org/bibliography/57641
Vahl, M. (1791). *Symbolae botanicae, sive plantarum, tam earum, quas in itinere, inprimis orientali, collegit Petrus Forskål,* ... Vol. 2. Copenhagen: Møller. https://www.biodiversitylibrary.org/bibliography/57641
Vahl, M. (1794). *Symbolae botanicae, sive plantarum, tam earum, quas in itinere, inprimis orientali, collegit Petrus Forskål, quam aliarum, recentius detectarum, exactiores descriptiones, nec non observationes circa quasdam plantas dudum cognitas,* ... Vol. 3. Copenhagen: Møller. https://www.biodiversitylibrary.org/bibliography/57641
Vahl, M. (1804). *Enumeratio plantarum, vel ab aliis, vel ab ipso observatarum cum earum differentiis specificis, synonymis selectis et descriptionibus succinctis.* Vol. 1. Copenhagen: Møller. https://www.biodiversitylibrary.org/bibliography/7564
Vahl, M. (1805). *Enumeratio plantarum.* Vol. 2. Copenhagen: Møller. https://www.biodiversitylibrary.org/item/31960
Verdcourt, B. (1988). A confusion in *Priva* (Verbenaceae). *Kew Bulletin* 43(4): 671–672. https://www.jstor.org/stable/4129965
Warming, E. (1891). Forsskål, Per. In: Bricka, C.F. (ed.), *Dansk Biografisk Leksikon.* Ed. 1, Vol. 5: 258–259. København: Gyldendal.
Wang, Q. [Qi] (2011). (1995) Proposal to conserve the name *Nilssonia* with that spelling (fossil Cycadopsida, Nilssoniaceae). *Taxon* 60(1): 240. https://onlinelibrary.wiley.com/doi/epdf/10.1002/tax.601031 & https://doi.org/10.1002/tax.601031
WFO (continuously updated). *World Flora Online.* https://wfoplantlist.org/ (accessed 1 Jul. 2024).
Wijk, R. van der (1959). Index Muscorum. Vol. 1 (A–C). *Regnum Vegetabile* 17: i–xxviii & 1–548.
Wilson, K.L. (2016). Report of the General Committee: 15. *Taxon* 65(5): 1150–1151. https://onlinelibrary.wiley.com/doi/epdf/10.12705/655.14 & http://dx.doi.org/10.12705/655.14
Wolff, T. (1980). Forsskål, Peter (Petter, Pehr, Petrus Forsskål). In: Cedergreen Bech, S. (ed.), *Dansk Biografisk Leksikon.* Ed. 3, Vol. 4: 494–495. Gyldendal.
Wolff, T. (1990). De zoologiske resultater af Den Arabiske Rejse. in: Rasmussen, S.T. (ed.) (1990). *Den Arabiske Rejse 1761–1767. En dansk ekspedition set i videnskabshistorisk perspektiv.* [The Arabian Voyage 1761–1767. A Danish Expedition seen through the History of the Sciences and Humanities.]: 231–237. København: Munksgaard.
Wood, J.R.I. (1997). *Handbook of the Yemen Flora.* Kew: Royal Botanic Gardens.
WoRMS (continuously updated). *WoRMS. World Register of Marine Species. An authoritative classification and catalogue of marine names.* https://www.marinespecies.org/ (accessed 10 Jul. 2024).
Wörz, A. (2007). The 'Botanische Reiseverein' – A 19th-century joint stock company for the collecting of herbarium specimens. *Huntia* 13(2): 121–141. https://www.huntbotanical.org/admin/uploads/03hibd-huntia-13-2-pp121–142.pdf
Yaron, I., Schiøtte, T. & Wium-Andersen, G. (1986). A review of molluscan taxa described by P. Forsskål and C. Niebuhr with citation of original descriptions, discussion of type-material available and selection of some lectotypes. *Steenstrupia* 12(10): 157–203. https://www.researchgate.net/publication/272161890
Zijlstra, G. (1998). Report of the Committee for Bryophyta: 4. *Taxon* 47(4): 859–862. https://onlinelibrary.wiley.com/doi/epdf/10.2307/1224193 & https://doi.org/10.2307/1224193

Index to Scientific Names

This index includes orthographic variants, corrected spellings, misidentifications, isonyms and homonyms of plants and animals mentioned in the text. The index does not distinguish names for animals and plants. Orthographic variants of the plant names in the original publications are in bold. A distinction between original spelling with capital or non-capital initial letters is maintained. Different author indications in the text for isonyms or homonyms are also maintained in this index, but validly and invalidly published and illegitimate and legitimate names are listed with the same typography, the conclusions about these distinctions can only be found in the discussion in the text. Names in current use are in italics but underlined. Names derived from Forsskål's surname and containing 'å' are alphabetized after all names with 'a' in the same position.